Praise for

THE SKEPTICS' GUIDE TO THE FUTURE

NAMED A BEST SCIENCE BOOK OF 2022 BY BARNES & NOBLE

"Steve Novella and crew are serious, knowledgeable nerds. They use their critical thinking skills to analyze what is and isn't possible... or likely. Along with some remarkable insights, they've written a little science-based fiction of their own that is a delight to read. It's a well-researched, well-reasoned reference book."

—Bill Nye, CEO, the Planetary Society

"[A]n entertaining evaluation of futurism...The result is pop science done right." ***—Publishers Weekly*, starred review**

"A gimlet-eyed look at the promises of technology and futurists past...An intriguing if bet-hedging work of futurology that calls into question the whole business of futurology itself." ***—Kirkus***

"A fun overview of both the current state of modern science and a general survey of the history of futurism." ***—Booklist***

"One aspect I liked about *The Skeptics' Guide to the Future* is how deep the Novella brothers dive into the science. This is not a book that shies away from deep technical and scientific explanations. The Novellas are serious folks, and their approach to projecting future developments is always grounded in established scientific theories and facts."

—Inside Higher Ed

"I think that anyone who has been fascinated with the future, as I assume most people have at one time or another, should check out *The Skeptics' Guide to the Future*... [I]t gives such an interesting take and view in which we should all look at the future. The realist perspective created more hope in me because it felt like the trajectory presented was achievable because of the building blocks already around."

—Cosmic Circus

"*The Skeptics' Guide to the Future* also explores the technology of our past and present—from biology to engineering, transportation to materials science, AI and quantum computing, and more—with the critical thinking skills the SGU crew have honed on their weekly podcast. The authors have crafted a text that's part James Burke's *Connections*, part where-we-stand today, and part sci-fi fandom, extrapolating in a science-based way to consider the future of technology, while also pouring cold fusion on ideas that will always defy the 'pesky laws of physics.'"

—AIPT Comics

Praise for

THE SKEPTICS' GUIDE TO THE UNIVERSE

"There are so many ways to be wrong, what we all need is a guidebook to being right. And here it is: *The Skeptics' Guide to the Universe* is an invaluable manual to avoiding all of the ways we can fool ourselves and be fooled by others. It's depressing to think of how many ways there are, but at least now we have no excuse for not being prepared."

—Sean Carroll, author of *The Big Picture: On the Origins of Life, Meaning, and the Universe Itself*

"A fantastic compendium of skeptical thinking and the perfect primer for anyone who wants to separate fact from fiction."

—Richard Wiseman, professor of the Public Understanding of Psychology at University of Hertfordshire and bestselling author of *59 Seconds*

"A lively, engaging, and very timely guide to navigating a world rife with misinformation and pseudoscience. This book will give you the tools to ferret out nonsense and confront your own biases—and hopefully change a few minds along the way."

—Jennifer Ouellette, author of *Me, Myself, and Why* and *The Calculus Diaries*

"Using examples ranging from Monty Python to Monty Hall, *The Skeptics' Guide to the Universe* offers the first ever page-turner that teaches you how to think clearly."

—Paul A. Offit, MD, author of *Bad Advice: Or Why Celebrities, Politicians, and Activists Aren't Your Best Source of Health Information*

Also by the Authors

The Skeptics' Guide to the Universe: How to Know What's Really Real in a World Increasingly Full of Fake

The SKEPTICS' GUIDE to the FUTURE

What Yesterday's Science and Science Fiction Tell Us About the World of Tomorrow

DR. STEVEN NOVELLA

with BOB NOVELLA & JAY NOVELLA

GCP

GRAND CENTRAL

NEW YORK BOSTON

Cover design by Jarrod Taylor
Cover illustration by David A. Hardy / Science Photo Library

Grand Central Publishing
Hachette Book Group
1290 Avenue of the Americas, New York, NY 10104
grandcentralpublishing.com
twitter.com/grandcentralpub

Originally published in hardcover and ebook by Grand Central Publishing in September 2022
First Trade Paperback Edition: September 2023

Grand Central Publishing is a division of Hachette Book Group, Inc. The Grand Central Publishing name and logo is a trademark of Hachette Book Group, Inc.

Library of Congress Cataloging-in-Publication Data
Names: Novella, Steven, author. | Novella, Bob, author. | Novella, Jay, author.
Title: The skeptics' guide to the future : what yesterday's science and science fiction tell us about the world of tomorrow / Steven Novella with Bob Novella and Jay Novella.
Description: First edition. | New York : Grand Central Publishing, 2022. | Includes bibliographical references and index.
Identifiers: LCCN 2022019456 | ISBN 9781538709542 (hardcover) | ISBN 9781538709566 (ebook)
Subjects: LCSH: Science—Popular works. | Science—Forecasting—Popular works. | Technological forecasting—Popular works.
Classification: LCC Q162 .N674 2022 | DDC 500—dc23/eng20220722
LC record available at https://lccn.loc.gov/202201945

ISBN: 9781538709542 (hardcover), 9781538709566 (ebook), 9781538709559 (trade paperback)

Printed in the United States of America

LSC-C

Printing 1, 2023

We dedicate this book to our father, who gave us the gift of a deep appreciation and fascination with all things science and technology, a love of science fiction, and a yearning for the wonders that the future might bring.

Contents

PART FIVE:
Science Fiction Tech—What Is and Isn't Possible

A Glimpse of the Future

Click, click.

Gerald stood in his closet alcove hitting the activation button but was rewarded only with a red light and an annoying click. Gerald could not be late. Not today. Then he noticed he forgot to set his dress type. He selected business attire for such an important day, and finally the light turned green. This time when he clicked the button, the robot arms came alive, dressing him in his favorite suit, tying his tie, and neatly combing his hair. He smiled wide and an ultraviolet beam of light cleaned his teeth. Perfect.

He strolled confidently out to the kitchen, where he was greeted by his wife, who was busy sending their kids off to school through the pneumatic tube system. They dropped away right before they could say goodbye to their dad.

"How are you doing? Nervous?" His wife managed a supportive look despite her obvious concern.

"It will be great," Gerald beamed, trying to convince himself as much as her. "This new line of autonomous robot servants is leaps ahead of the Model 2s. They practically sell themselves."

As if to emphasize his point, Giles, their Model 2 home butler, clanked over to Gerald, hat in hand, and said in a tinny voice, "Morning, sir. Shall I call your car?"

"Of course, Giles. Look at the time. It should be here already."

Lights flashed on Giles's "face," indicating he was engaged with the home central computer while his metallic frame stood motionless. Gerald grabbed his hat from the steel fingers, visibly losing his patience just a little.

"Don't forget your breakfast, dear." His wife pointed to his plate at the table, on which sat two small capsules. "You need your energy for the meeting."

He downed his breakfast, placed his hat carefully on his head, kissed his wife, and then strode through the shimmering rectangle of energy that served as the front door of his home. The thorium nuclear engines of his car were purring nicely as the vehicle floated in front of him, the driver's door opening automatically as he approached.

In the twenty minutes it would take him to fly to work, Gerald reviewed his pitch. He was the chief engineer on this project and had been over the prototype Model 3 with a fine-toothed comb. The higher-ups were sure to fund production, but everything still had to go perfectly.

Azra removed her VR (virtual reality) goggles, the retro-future world vanishing around her, replaced by her somewhat stark office.

"I can't take any more. Is that really what they thought the future was going to be like back then? And what's with the creepy 1950s vibe?"

The swarm of nanites that made up Briar's avatar laughed. That was about the reaction they were expecting. "I know, it's great, right? You gotta love the robot butler. So retro."

Briar's smile was just a bit delayed, leading Azra to suspect she was, yet again, talking to their AI simulacrum rather than the person themselves. "Listen, the Museum of the Future wants this exhibit to go live in two days. I need to put the finishing touches on the project and upload it by tomorrow. I really need your feedback."

"Well..." Azra wasn't sure how to put this. "The butler is kinda funny, if that's what you're going for. But why is everything so... analog? Where's the embedded AI, the virtual overlays? It all looks a bit cartoonish."

Briar laughed again. "Hey, I'm just reconstructing their vision of the future. I didn't make it up. Just watch the rest and then ping me with your feedback. Remember, you're not critiquing their futurism. I just want to know how the VR looks and feels."

Azra gave Briar a curt nod, then dismissed their avatar with a wave. She reached for the VR goggles, ready to immerse herself again in a world that never existed or will exist, except in the misguided imagination of someone long dead.

PART ONE

An Introduction to the Future

1. Futurism—Days of Future Passed

The future begins with the past.

The future is a wild fantasy. It's feverishly concocted out of our hopes, fears, biases, ignorance, and imagination, saying far more about us than what is to come. Predictions of the future are really just reflections of the present. And that means we're really bad at predicting what the future will bring. But that's not going to stop us from trying to do just that—it's simply too irresistible.

We can, however, try to learn from futurism's checkered past, correct what errors we can find, and perhaps do a little better. Along the way, we can learn about the past and present of the technologies that dominate our world. We can follow the arc of the history of science and technology and perhaps extrapolate it a little bit into the future. My brothers and I have been doing this our whole lives.

As children growing up in the '60s and '70s, we were in love with science, technology, science fiction, and the incredible promise of the future. We were too young to have experienced the disappointment of future promises repeatedly failed, and so we naively believed the advances set before us. Many of them are now solid clichés about the future, but back then we anxiously anticipated our flying cars, jetpacks, moon settlements, and intelligent robot servants.

Our fondness for science fiction didn't help. The movies and television we watched depicted a near future with technology that now seems to be about a century premature. In *The Six Million Dollar Man* Steve Austin sported prosthetic robotic limbs that fifty years later are still not even close to achievable. In *2001* we were supposed to have space stations and sentient computers. And weren't researchers working on beaming vivid experiences directly into our brains? Even dark futures, like the year 2019 depicted in *Blade Runner*, had flying cars and genetically modified androids indistinguishable from humans. No matter how socially and environmentally devastated the future was presented to be, what I marveled at was the technology. We could work the other stuff out—as long as there were flying cars.

Our techno-optimism was likely significantly influenced by growing up in the era of Apollo. We were landing people on the moon and using "advanced" computers, and despite a few hiccups, it all worked out well. Watching Gene Cernan step off the moon and back into the lunar lander during the Apollo 17 mission in 1972, my younger self couldn't conceive that fifty years later we still would not have returned, let alone not have a settlement on the moon. Where is Moonbase Alpha?

The flip side of this disappointment and false promises is that some of the biggest technological advances in the last half century, the ones that have had the most significant impact on our lives, were not featured in future predictions or science fiction. As I write this, I carry in my pocket a supercomputer (by the standards of my youth) that allows me to communicate instantly through video, audio, or text to almost anyone, anywhere in the world. As a bonus, I have access to my entire personal music library, it serves as a digital camera that can take as many photos as I want without the need for film, and can even give me directions to anywhere I want to go. It can access practically the sum of human knowledge in a searchable interactive format. If I get bored, this device can play movies and contains countless video games that would have blown child-me away.

The smartphone and the World Wide Web that can be accessed through it, along with social media, online shopping, countless apps, and other features, are absolute miracles of future technology.

It far exceeds what I would have thought possible thirty to forty years ago. Past depictions of the future generally did not anticipate anything like it. Even *Star Trek*, a favorite techno-optimist utopian future, did not see our digital revolution coming.

So, we have made great technological strides in the last fifty years, just not in the ways that we thought. Why is it that people are consistently so bad at predicting the future? If we can understand this, perhaps we can do it a little better. Or perhaps the forces that shape the future are too chaotic to predict with any accuracy beyond a certain point, like trying to predict the weather.

But while we cannot predict specific weather with much accuracy, we can better predict overall changes in climate. It's easier to predict that travel will get faster in the future, for example, rather than the specifics of automotive technology. One potential fix to bad futurism, therefore, is to focus on large trends, rather than trying to imagine tiny details. Even there, however, futurists can get tripped up.

In the movie *Minority Report*, for example, they presented a thoughtful picture of the near future of 2054. I will not be able to say how accurate it was for another thirty years, but one choice stuck out to me: People were using teeny-tiny phones. The movie was made in 2002, prior to smartphones. The trend then for cell phones was to get smaller and smaller, so the writers extrapolated this out another fifty years.

Unfortunately for them, the iPhone was released in 2007, essentially reversing the trend by fundamentally changing how personal phones are used. Suddenly, screen real estate became a premium, and inevitably phones got larger. The iPhone was a disruptive technology, changing the status quo of an entire industry and changing our lives forever. Perhaps we are now settling into a range of screen sizes that represents the optimal balance of portability and usability, depending on some personal and situational variables, or maybe a new disruption will occur.

Companies are indeed looking for ways to disrupt the cell phone market again by creating foldable or expandable phones reminiscent of a time before iPhones. Will this technology take off, remain niche, or completely fail? If you could reliably predict even the next step in how a single technology will evolve, you would be a tech millionaire. Even scientists and tech leaders in the past famously made terrible predictions about the future, including about technology in which they were involved. Take Thomas Edison, who said, "The phonograph has no commercial value at all" in the 1880s. Or Ken Olsen who claimed in 1977 that "there is no reason anyone would want a computer in their home." Even if you can foresee that first step, try predicting the next fifty steps for a thousand technologies. That's the future.

Of course, "the future" includes one second from the moment you read these words, to the ultimate heat death of the universe in 10^100 years (some trends are inevitable). Different factors apply depending on how far into the future you are trying to predict. The near future, say ten to twenty years, can benefit from high-probability extrapolation of existing trends, as well as coming technology that is already in the works. The medium future, twenty to a hundred years, gets a lot harder, but if you focus on the big picture and give yourself some wiggle room, you might get a glimpse of life in a century.

The far future, more than a hundred years, is where things really get interesting, and the technologies we are now just beginning to explore reach their full mature potential. While it may be easy to predict that some technologies will eventually be realized, the wiggle room for predictions here is in how long it will take. We may not be able to envisage when essentially complete brain-machine interfaces will exist, but when they do, we can imagine what it may be like. The far future is where we can enjoy speculating about entirely new technologies that are now only a footnote in a physics paper about some newly discovered aspect of nature.

Our guide through the future will cover all of this—advancements in existing technologies, exploring emerging technologies, and speculating about fantastical tech from possible futures. As much as possible, science will be our guide.

Through it all, we will maintain a highly skeptical eye because that's what we do. In addition to being science enthusiasts and technophiles, we are also scientific skeptics. For the last quarter century, we have been studying and promoting critical thinking and scientific literacy. We host the award-winning podcast *The Skeptics' Guide to the Universe*, and our first book, of the same name, is a primer on science and critical thinking.

This means we always try to temper our enthusiasm for the future with sharp criticism. We let the failures and disappointments of the past inform our thoughts about the future. It is not enough, however, to simply be cynical. Being skeptical means separating the probable from the improbable, with solid evidence and logic.

Sometimes we let our enthusiasm get the better of us, but in the end, we always bring it back down to reality. This is, after all, a skeptic's guide to the future.

The future, ironically, begins in the past. Our journey starts with the history of futurism to see what it can teach us.

2. A Brief History of the Future

The pitfalls of futurism.

Yogi Berra famously said, "The future ain't what it used to be." (You knew that quote would make an appearance, right? Although he wasn't the first person to utter this phrase.) He had a quirky genius for making a concept that ultimately makes sense sound self-contradictory or nonsensical. What has changed is not necessarily the future itself, but our beliefs about the future. Futurism ain't what it used to be.

In traveling back in time and into the future at the same time, we can look at some of the visions of yesterday's future—what did they get right and what did they get spectacularly wrong? When tracking the errors of past futurists, some common themes emerge—what I call "futurism fallacies"—and they will help us shape our own future vision.

Futurism Fallacy #1—Overestimating short-term progress while underestimating long-term progress.

One core challenge of predicting the future is trying to determine not only what technology will develop, but also how long it will take. It is almost a certainty that eventually we will develop fully sapient general artificial intelligence. What is extremely challenging is predicting when we will cross that finish line. When trying to glimpse the future, there is also a tendency to overestimate short-term advancement, while underestimating long-term advancement. We see this frequently in science fiction movies—whether they are serious, comedic, utopian, or dark, the technology in twenty to thirty years is typically portrayed as transformational, rather than more realistically incremental. So, from *Back to the Future* to *Blade Runner*, we have flying cars by 2015.

This overestimation of short-term progress comes partly from the tendency to think of "the future" as one homogenous time, much as we often think of "the past" as one indistinct era. My favorite example of this is that Cleopatra (69–30 BCE) lived closer in time to the Space Shuttle (first launch in 1981) than the building of the Pyramids of Giza (2550–2490 BCE).

But, getting back to the future (heh!), we often imagine that "the future" contains whatever the next big advance is expected to be in any given technology. So instead of using phones, we are using video phones in the future. Instead of driving regular cars, we are driving electric, self-driving, or flying cars. We fly to other continents in commercial jets now, so in the future we must be flying rockets instead. Even if "the future" is only twenty years from now, we imagine all these technological transformations have already taken place, therefore overestimating short-term progress.

Underestimating long-term progress is mostly a matter of simple math, because technological progress is often geometric rather than linear. Geometric progress means doubling (or some other multiplier) every time interval, so progress looks like 2, 4, 8, 16, 32, while linear means simple addition every time interval: 1, 2, 3, 4, 5, and so on. You can see how geometric is much faster than linear, especially over the long-term. The best example of this is computer technology, such as hard-drive capacity and processor speeds. Processor speeds have roughly doubled every eighteen months, so in the past forty-five years, processors have not become several times faster—they have become millions of times faster. Underestimation also occurs because game-changing new technologies are often missed.

However, while there is a general tendency to overestimate short-term and underestimate long-term progress, each technology follows its own pattern of progress. Therefore, it can be difficult to pick technological winners and losers. The problems we are trying to solve with technology may also be nonlinear, getting progressively harder to advance at the same rate. At first, we may pick the low-hanging fruit and progress may be rapid, but then further advances become increasingly difficult, with diminishing returns and perhaps even roadblocks. If we project early progress indefinitely, that will result in overestimating the strides we will make. But then geometric advances and game-changing innovations eventually catch up, causing us to underestimate long-term progress.

A humorous showcase of this fallacy is the short film produced by General Motors in 1956, which imagined the "modern driver" of 1976. They get everything wrong. The film was made to promote their gas turbine engine technology—remember those? Probably not, because they never came into wide use. Several car companies, most aggressively Chrysler, tried to develop a gas turbine engine to replace the internal combustion engine, and none succeeded.

The film also featured "auto control" in which the car was able to take over steering from the driver—in 1976, remember, about a half century too early. But in order to do this, the driver had to first enter the "electronic control lane" and then synchronize the car's velocity and direction with the external control. This was all done with the help of people in control towers that lined the highway through radio communication.

Science fiction depictions of the future are also rife with this fallacy. In 1968, the movie *2001: A Space Odyssey* chronicled a mission to Jupiter (still beyond our current technology), with crew members in cryosleep (also not possible) and featuring a fully artificially intelligent computer, the HAL 9000. These technologies are all at least fifty to a hundred years premature.

Professional futurists, such as Isaac Asimov, frequently fell for this fallacy. In 1964, he made predictions for 2014, fifty years in the future, for the world's fair. His forecasts were published in the *New*

York Times, although in fairness, he admits these are only "guesses." He predicted:

> It will be such computers, much miniaturized, that will serve as the "brains" of robots. In fact, the I.B.M. building at the 2014 World's Fair may have, as one of its prime exhibits, a robot housemaid—large, clumsy, slow-moving but capable of general picking-up, arranging, cleaning and manipulation of various appliances. It will undoubtedly amuse the fairgoers to scatter debris over the floor in order to see the robot lumberingly remove it and classify it into "throw away" and "set aside." (Robots for gardening work will also have made their appearance.)

What about energy?

> And experimental fusion-power plant or two will already exist in 2014. (Even today, a small but genuine fusion explosion is demonstrated at frequent intervals in the G.E. exhibit at the 1964 fair.) Large solar-power stations will also be in operation in a number of desert and semi-desert areas—Arizona, the Negev, Kazakhstan. In the more crowded, but cloudy and smoggy areas, solar power will be less practical. An exhibit at the 2014 fair will show models of power stations in space, collecting sunlight by means of huge parabolic focusing devices and radiating the energy thus collected down to earth.

Again, these predictions are at least a half-century premature. In general, futurists need to be much more conservative in their estimates of short-term advancement. It seems like doubling or even tripling the estimated timeline is a reasonable rule of thumb. Anticipate roadblocks, blind alleys, and troubling hurdles, and your estimates will likely be closer to the mark.

Futurism Fallacy #2—Underestimating the degree to which past and current technology persists into the future. Corollary—assuming we will do things differently just because we can.

One particularly ambitious look forward was the 1967 film by the Philco-Ford Corporation imagining the world of 1999, starring a young Wink Martindale. Those thirty-two years were full of advancements with the advent of personal computers, the internet, and essentially a transition to digital technology.

The authors of this film could not see through the thick veil of these technological revolutions, so they relied heavily on their own hidden assumptions, falling for many of the futurism fallacies. They assumed that many aspects of daily life would change simply because future technology would allow for such change. Everything thirty-two years in the future has to be different, right? History has shown, however, that past technology persists into the future to an incredible degree.

In their depiction of a typical day in 1999, even the simple act of drying one's hands at home had to be done using the most advanced technology possible, such as infrared lights and air blowers. Drying your hands on a towel seems too old school for "the future." Communicating from the next room was done through video. In this future, all food is stored frozen in individualized portions and heated up in minutes by microwave to serve, replacing all cooking, except perhaps for special occasions. The central computer monitors nutritional and caloric needs and suggests the appropriate menu.

Asimov made similar predictions about cooking in the future (again from his 1964 world's fair predictions):

> Gadgetry will continue to relieve mankind of tedious jobs. Kitchen units will be devised that will prepare "automeals," heating water and converting it to coffee; toasting bread; frying, poaching or scrambling eggs, grilling bacon, and so on. Breakfasts will be "ordered" the

> night before to be ready by a specified hour the next morning. Complete lunches and dinners, with the food semiprepared, will be stored in the freezer until ready for processing. I suspect, though, that even in 2014 it will still be advisable to have a small corner in the kitchen unit where the more individual meals can be prepared by hand, especially when company is coming.

The idea that we will be utilizing essentially the same culinary techniques in the future, despite the fact that we've been cooking food over heat for millennia, just doesn't fit a futurist lens. But the reality is we still buy raw vegetables and cut them with knives on a wooden cutting board, and then steam them or stew them in a pot. My modern kitchen and the process I use to cook would be fully recognizable to someone from fifty, even a hundred, years ago depending on the recipe. In fact, I recently purchased some hand-forged kitchen knives. Sure, the appliances are all more efficient, and may have had some incremental functional improvements, but mostly they are the same. The biggest innovation is the microwave oven, which I, like most, use for heating, not for cooking.

Sometimes we do things the old-fashioned way because we want to, or because the simple way is already pretty close to optimal. Sometimes convenience isn't the most important factor (that is another lazy assumption about the future—everything is about optimizing convenience). Recently, several people close to me purchased the automatic coffee makers where each cup is made from a prepackaged individual container of ground coffee. This method prioritizes ease, speed, and convenience, and these coffee makers became very popular. But after a couple of years there was a backlash, and all it took was being exposed again to a really well-brewed cup of coffee. Compounded with environmental concerns over the waste of all the plastic used in those individualized packages, suddenly the swill they were drinking out of convenience simply wasn't good enough.

Now many of them (I don't drink coffee, so I watched this play out from the sidelines) have swung back to the other end of the spectrum, prioritizing quality. They grind their beans fresh and may go through an elaborate process such as slowly pouring boiling water over those grounds in search of the perfect cup of coffee. They enjoy the ritual, and it builds their anticipation of flavorful enjoyment.

Old technology can be remarkably persistent. We still burn coal for energy. Our world is still largely made out of wood, stone, steel, ceramic, and concrete—all materials that have been used for thousands of years. Plastic is probably the one new material that has shaped our modern world, but not everything is made from plastic just because it can be.

None of this is to downplay the truly transformational technologies that make up our modern world and have changed our lives. But the future always seems to be a complex blend of the new and the old. The trick is predicting which things will change, and which will substantially stay the same.

Futurism Fallacy #3—Assuming there is one pattern of technological change or adoption. Rather, the future will be multifaceted.

Sometimes new technologies completely fail and fade away (like the gas turbine engine), sometimes they are adopted but fill a smaller niche than initially assumed (like microwave ovens), and sometimes they completely replace (except for nostalgic or historical purposes) the previous technology (like cars did to the horse and buggy). There is no one pattern.

We must also recognize that there are many competing concerns, and this is often why it is extremely difficult to predict how a new technology will play out. Convenience is not everything, and we do not widely adopt new technologies just because they are new. In addition, there are considerations of cost, quality, durability, aesthetics, fashion, culture, safety, and environmental effects. Even the concept of convenience itself can be multifaceted.

What tends to happen is that many technologies exist in parallel, each finding their niche where these combinations of factors make the most sense. I am writing this book on my computer, but sometimes I take notes by writing them down on a piece of paper. It's just more convenient for some applications. Sometimes I listen to books on audio, sometimes I read them on my ebook, and sometimes I like the feel of a physical book in my hands.

We still use natural wood in home construction because of its cost and how easy it is to work with, and for a desired aesthetic. In fact, antiques have a high value partly for their rustic or quaint appearance in home interiors. Conversely, I may spring for artificial wood for my deck because of its weather resistance and lower maintenance.

I drive to work in a car that would mostly seem ordinary to a driver from the 1950s, but they would likely be blown away by my GPS and entertainment system. And yet, I still usually just listen to news on the radio.

Futurism Fallacy #4—Anticipating the end of history.

In his book, *Predicting the Future*, Nicholas Rescher points to a tendency to assume "the end of history"—that society reaches its equilibrium point, and once achieved we have endless peace and prosperity. In this utopian future, not just convenience but also leisure become everything. Well, history doesn't stop, at least it hasn't so far.

A good example of this fallacy is the 1920s film about the twenty-first century *Looking Forward to the Future*, set after the "War to end all wars" where there would be never-ending peace and prosperity, with increasing leisure time. It predicts people wearing electric belts to control their climate. Men (not women) would be outfitted with a utility belt that would contain, "telephone, radio, and containers for coins, keys, and candy for cuties." Planes would be enormous, designed like luxury cruise ships, with lounges, dining areas, and activities. The farther back in time we go, the more outlandish our present "future" becomes.

This fallacy mostly stems from a lack of imagination—thinking that all of our current problems will be solved by technological advancements. Once that happens, we will have achieved a stable utopia. But what always has happened so far is that as we solve one problem, new problems emerge. Even the technology we develop to make our lives better can come with a suite of its own challenges—new resources become precious, power shifts, and new conflicts arise. When past futurists looked at the advent of the internal combustion engine, they did not imagine the challenges of global warming, or the rise of power centers in the deserts of the Middle East.

History does not end; it just keeps churning.

Futurism Fallacy #5—Extrapolating current motivations and priorities into the future. Corollary—it's still not all about leisure and convenience.

This assumption of increased leisure time was not unreasonable a century ago, as the industrial revolution really did free developed societies from previous crushing drudgery. Machines taking over many of the worst repetitive, time-consuming, and dangerous tasks was a defining feature of that era. It stands to reason that they would extrapolate that trend into the future, so they did. This assumes that current trends will continue indefinitely into the future, but they rarely do.

In the United States, for example, the forty-hour workweek did not come about because of technological advancement. In fact, industrial factories made workers more productive, and therefore their hours of work more valuable. Achievement of a forty-hour workweek was the result of a labor fight, fought over a century, and finally achieved by federal law in 1940. Since then, the workweek had been stable but has been increasing recently with the rise of new forms of contract work that fall outside these regulations.

A 2014 national Gallup poll put the average US workweek at forty-seven hours. This number is higher for competitive industries, or gig workers like Uber drivers, who might work a hundred hours a week. This increase is ironically driven by modern technology that facilitates contract work, which would have been hard to predict a hundred years ago. Now there is a push for a shorter workweek and increased work from home, made possible by computers and remote conferencing. These trends may have had a little boost from a once-in-a-century pandemic that was difficult to predict.

The core problem with this fallacy is unrecognized and often lazy assumptions. In the past, progress was equated with convenience, and so future progress was seen through that narrow lens. The question is—what hidden assumptions are we making today that color our thinking about the future?

Futurism Fallacy #6—Placing current people and culture into the future.

"The past is a foreign country: they do things differently there" are the first words of L. P. Hartley's *The Go-Between*. By extension, the future is a foreign country too. When we envision the future, by default we tend to place people like ourselves in that future. But this would be the equivalent of imagining people from the Middle Ages living in the twenty-first century. One thing is certain—the people of the future will not simply be us with better technology. People and culture change, often in ways difficult to predict.

People in future societies are likely to have different priorities, ethics, and a different relationship with their technology. We can see this today, as often parents don't understand how their own children use the latest social media app. We tend to accept technological changes over time, and eventually the unthinkable becomes every day. We may be squeamish about genetic manipulation, for example, but people in a century may think nothing of it. How will individuals in future generations relate to artificially intelligent robots?

This fallacy is common in science fiction, which is why the classic 1956 film *Forbidden Planet* imagined a starship from the twenty-third century being crewed entirely by men who would have seemed completely at home aboard a World War II battleship. Without more imagination, their vision of the future is already rendered obsolete.

This futurism fallacy is so common it has become almost synonymous with retro-futurism. We can see this, for example, as a deliberate aesthetic in the *Fallout* video games. It presents the world of 2071 as imagined by futurists in the 1950s, complete with unaltered 1950s culture, fashion, and attitudes. But these people from the 1950s are living in a world of retro-futuristic technology, including robots and nuclear-powered cars.

People and technology evolve together as a dynamic system, and yet futurists tend to imagine that time stands still, except for technology.

Futurism Fallacy #7—Everything that happens in the future will be planned and deliberate.

One more common assumption that shines through these older attempts at predicting the future is that society and technology in the future will be more planned, designed, deliberate, and controlled. The 1935 *City of the Future* film makes this assumption explicit, stating outright that in the future every detail will be planned. This high level of planning fits better with utopian rather than dystopian futures, but it is based on the notion that the powers that be will craft our future to best meet our needs.

In the Ford film about 1999 (the one with Wink Martindale), the center of their futuristic home is an entire room dedicated to the home's central computer. The computer itself looks like a monstrosity out of the 1950s, with banks of switches, blinking lights, and vacuum tubes. It is presented as an overmind that controls every aspect of the home and the family's life, from planning meals to homeschooling the children.

Reality, of course, remains far messier. The deep philosophical principle that shit happens continues to hold true. Things evolve in a chaotic way, for often quirky and highly contingent reasons, again contributing to our inability to accurately predict the future.

The Segway promised to change the way people get around. This is still sexy technology: an electric two-wheel stand-up platform that can transport you quickly around a city or large indoor space like a mall. Steve Jobs reportedly said that the Segway will be "as big a deal as the PC." However, the Segway just never caught on, largely for practical and economic reasons. Perhaps the hardest thing to predict about technology is how people respond to and use that technology—will people want to ride wheeled platforms around the city? Is it better than other options, enough to justify the expense?

Similarly, futurists failed to ask, will people want to communicate routinely with video? The surprising answer turned out to be no; they would rather text, of all things. I don't think anyone predicted this. Video phones are almost ubiquitous in portrayals of the future, but they remain a tiny slice of how we communicate today, despite the technology being fully mature.

Our utilization of future tech is likely to contain a lot of chaos because people are unpredictable.

Futurism Fallacy #8 (the Steampunk Fallacy)—Extrapolating existing technology into the future without considering new and potentially disruptive technology.

Another feature of past predictions of the future is what is not there, the things they did not predict. Looking back from the vantage point of the 2020s, the biggest technological advance over the last few decades, largely missed by futurists, is the digital revolution. Past portrayals of today are often still analog—they extrapolated their current technology into more advanced forms but failed to account for the possibility of disruptive technologies.

One dramatic example of this is the epic science fiction series by Isaac Asimov, the Foundation trilogy, written in the 1940s and 1950s. He imagines a distant future, thousands of years from now, that is completely analog. In the first two books, computers are not even mentioned. It is also striking how much the people of his future resemble those from the 1950s, complete with hat-wearing, cigar-smoking, male domination.

This fallacy often reminds me of steampunk fiction, which aesthetically imagines a world in which steam industrial technology remained the cutting edge and continued to advance to more complex and intricate instruments and devices. In this world, it was never replaced by electronic or digital technology.

We keep imagining a steampunk future. Futurists of the early nineteenth century missed electronics, and those of the first half of the twentieth century missed computers, and later missed the miniaturization and therefore wide distribution of computers. Typically, we don't start seeing new technologies in future fiction until it already exists.

Even then, we are terrible at picking technological winners and losers. Remember the coming hydrogen economy? In the early 2000s, it was widely believed that internal combustion engines would be replaced by hydrogen fuel cells in vehicles, and this would be the cornerstone of a hydrogen-based economy. Hydrogen fuel cell vehicles are not dead yet, but they are soundly losing to battery-electric vehicles, which are clearly the favorite to replace gasoline engines (they are just more energy efficient than hydrogen).

Back to Wink Martindale in 1999; the filmmakers do anticipate shopping from home, but it's all analog—a camera pans across items for purchase. And of course, the wife makes purchases in one room, while the husband pays for it at his bank of consoles in the next, complete with knobs and buttons, but no keyboard or other interface. They anticipated doing something in a new way but no further than an extrapolation of then-current technology.

Forecasting which new technologies will be game changers and which will fizzle is perhaps the greatest challenge of futurism. This fallacy also produces the greatest fails because a disruptive technology (or its absence) can completely change our vision of the future.

Futurism Fallacy #9—Assuming the objectively superior technology will always win out.

Yet another challenge of predicting the future is competitive technologies. Around the turn of the twentieth century, there was intense competition among steam-powered, electric, and gasoline-powered cars. Whichever technology won out would materially shape the next century and beyond in many ways. Of course, in hindsight it seems inevitable that gasoline engines would win, but it wasn't. Each technology had its advantages and disadvantages.

While this is a complex story, it did mostly come down to infrastructure. Electrification was not yet sufficient to allow convenient intercity travel with electric cars, and the availability to frequently refill the steam engines with water was also a limiting factor. Gasoline simply beat the others to developing the critical infrastructure, which created a self-reinforcing feedback. Adoption led to further investment in infrastructure, which led to further use. It also helped that Ford chose the gasoline engine for his first mass-produced car—a choice by a single individual that could have changed the technology landscape for an entire industry.

This was not the only historical competition where the winner was not inevitable. For long-distance travel, jets now seem obvious, but rockets were seriously considered. In fact, Elon Musk's SpaceX plans to bring back the idea of rockets for the longest trips. They certainly would be much faster.

Remember General Motors' gas-turbine engine? These are quieter, smaller, less polluting, and run cooler than internal combustion. They also started more reliably in cold environments than cars of the time did. But they were less fuel efficient and more expensive to produce. Never able to be cost competitive, they failed to become popular.

Often the marketplace victory of VHS over Betamax for home video recording is given as the classic example of this phenomenon, as Betamax was seen as a superior technology because of higher resolution, better sound, and a more stable image, and so it grabbed the early market. However, the makers of Betamax got one decision wrong—their more compact tapes could only record one hour, while VHS could record two or more. A typical movie could be recorded onto a single VHS tape, and that one convenience is what likely doomed Betamax.

Ultimately, what makes a technology "superior" may be subjective.

Futurism Fallacy #10—Failing to consider how technology will affect people, our options, and the choices we make.

In 1981, Bill Gates, cofounder of software giant Microsoft, said, "No one will need more than 637 KB of memory for a personal computer—640 KB ought to be enough for anybody." The computer on which I am writing these words has more than 32 million times as much memory. Many very smart people in the past have all fallen for one or more of the futurism fallacies I have been detailing in this chapter. Even still, the Bill Gates quote is shocking in hindsight. He seemed to miss the fact that once the personal computer became popular, people would want to do more with their device. That desire would benefit from more memory and power, which would then allow for other applications that in turn needed yet more memory and power. It can be reasonably argued that the computing power-hungry gaming industry is driving computer technology, something Gates apparently did not anticipate.

How inevitable were the technologies we now take for granted? It was once possible that we would be living in a world today powered by direct current and fueled mostly by nuclear power, in which all-electric vehicles have always been standard, long-distance intercity travel is mostly by rockets, and our home devices are powered by nuclear isotope batteries. Was the personal computer really inevitable? What if it remained largely a device marketed to corporations and institutions for big computing jobs? If no personal computers, then there would be no World Wide Web, no social media, no smartphones. How different would our world be?

History is full of singular events that changed the course of technology. If not for the Hindenburg explosion, might zeppelins still be a popular form of travel? It seems likely they would eventually have been replaced by commercial jets, but perhaps they would have survived the way ocean cruise lines have, as luxury vacations rather than a means of getting to a destination.

After the near disaster of Apollo 13, in which a small onboard explosion threatened the lives of the crew and forced them to return to Earth without ever landing on the moon, the last two Apollo missions, 18 and 19, were canceled. There were certainly other factors involved as well, but a different set of circumstances might have led the United States to complete the Apollo program and continue plans for crewed missions to the moon and even Mars. Our space program might be in a very different place today were it not for the one short circuit.

If our present was not inevitable, but rather the quirky creation of many individual choices and events, then no particular future is inevitable either. The future will be indelibly shaped by the choices we make today. These choices are made at every level, from individual purchasing decisions to the judgments of CEOs at large corporations. How we invest in infrastructure will also matter—do we invest billions in the information superhighway? Do we invest in charging stations or hydrogen refueling stations?

Quirky decisions by people, individually and collectively, mold the future in unpredictable ways. But there is another layer here as well—culture and society. In the next chapter we will explore how societies of the future are likely to be different, and how that impacts our attempts at predicting the future of technology.

3. The Science of Futurism

People of the future will not be the people of today.

If we are going to apply a truly skeptical filter to our exploration of the future, that means we must base our musings as much as possible on science, empirical evidence, and logic. In his excellent Foundation series mentioned previously, Isaac Asimov writes about the science of "psychohistory" in which these cultural trends are studied scientifically so that the future course of human history can be mapped. Even in Asimov's optimistic fiction, this process has to be constantly monitored and revised, and even then, fails spectacularly because of the introduction of unpredictable quirky elements.

Is "psychohistory" even a theoretical possibility?

Opinions as to whether there can even be a scientific approach to the future range the entire spectrum. In his 2003 book, *Foundations of Future Studies*, Wendell Bell argues that futurism as an academic field represents "a body of sound and coherent thought and empirical results." He therefore thinks futurism is a legitimate field of academic study. However, despite his optimism, the field of futurism has been waning in academia.

Futurism might be struggling to win academic acceptance because of the attitude of its many detractors. At the other end of the spectrum, for example, is William Sherden's 1998 book, *The Fortune Sellers: The Big Business of Buying and Selling Predictions*, in which he likens futurism to astrology. That is pretty harsh, comparing legitimate attempts to map the likely course of future science, technology, and society with an ancient superstition. But Sherden has a point when it comes to their relative success rate.

Is there, then, any scientific analysis we can bring to bear when thinking about the future? Identifying where past futurists have gone wrong and then trying to make error corrections, as we will try to do here, is certainly one reasonable approach. But how about looking at the progress of technology through history? Are there any big trends that we can reasonably extrapolate into the future?

In his 1999 book *The Age of Spiritual Machines*, Ray Kurzweil proposed the "law of accelerating returns." He argues that in any evolutionary system, such as technological progress, over time there will be an exponential increase in the rate of progress. For example, efficiency or speed or power will double over some interval of time. The most obvious recent example of this is Moore's law: Computer scientist Gordon Moore observed in 1965 that the number of transistors per silicon chip doubles every eighteen to twenty-four months. This trend has continued fairly consistently since then. Now we all enjoy the fruits of the exponential growth in computer drive capacity and processing speed. But is this pattern typical?

Kurzweil argues explicitly that it is if you take a broad view of evolutionary progress. He states:

> An analysis of the history of technology shows that technological change is exponential, contrary to the common-sense "intuitive linear" view. So we won't experience 100 years of progress in the 21st century—it will be more like 20,000 years of progress (at today's rate).

This view is not universally accepted, however. Sometimes technology runs into hurdles that it cannot overcome. That's why we still don't have that flying car—it takes a lot of energy to keep something heavy enough to carry people in the sky and rolling on the ground will always be more energy efficient.

While we cannot make this sort of prediction for each type of technology or application, it does seem like a reasonable reading of history that technological progress is accelerating and probably will continue to do so. But here too we cannot extrapolate current trends endlessly into the future. Perhaps we will run into some general technological barriers that will stall overall progress for extended periods of time. Those pesky laws of physics may impose limits that require discovering new laws or developing entirely new technologies to overcome, and that could take an unpredictable amount of time.

The challenge is that we are trying to get an overview of the arc of technological progress while we are still in the middle of it. We also only have one data point: humanity. It would be fascinating to have access to a galactic database in which we could review and compare the technological histories of dozens of civilizations and look for patterns. But that's not likely to happen anytime soon, so we must make do.

Many futurists approach technological advancement over time from an evolutionary perspective but draw different lessons from evolution than Kurzweil. In a 2019 paper, Mario Coccio writes that technological advancement is a complex system, involving technical choices, technical requirements, and scientific advances that attempt to solve complex problems. He references two underlying theories: One is the theory of competitive substitution, in which better technologies replace older ones. However, Coccio also introduces the idea of technological parasitism, which looks at technological advancement as a complex system of interacting technologies.

For example, as car engines become more powerful, this drives innovation to optimize car tires, suspension, and steering. Overall improvement in cars then affects how people use them, impacting society in unpredictable ways, such as where people live and how they plan their travel. These changes, in turn, drive further transformations in cars and other technology.

Clearly predicting how complex interacting systems will behave is extremely difficult. But not all past predictions were laughably wrong. There were some impressive hits if you consider the broad picture. Mark Twain of all people, in a way, predicted the internet. In his 1898 short story "From the 'London Times' in 1904," he wrote:

> As soon as the Paris contract released the telelectroscope, it was delivered to public use, and was soon connected with the telephonic systems of the whole world. The improved "limitless-distance" telephone was presently introduced and the daily doings of the globe made visible to everybody, and audibly discussable too, by witnesses separated by any number of leagues.

In 1900, engineer John Elfreth Watkins made a number of predictions that anticipated digital color photography, cell phones, tanks, and television. He wrote, for example, "Photographs will be telegraphed from any distance. If there be a battle in China a hundred years hence snapshots of its most striking events will be published in the newspapers an hour later. Photographs will reproduce all of Nature's colors." Sure, he also got a lot wrong, like the disappearance from the alphabet of the letters *C*, *X*, and *Q*, but his technology predictions were prophetic.

Futurists envisaged the widespread use of refrigerated cars for transporting produce, which has had a significant impact on the modern diet. Mobile homes, birth control, and medical imaging were also predicted (at least as general concepts).

In a 1967 report, Walter Cronkite gave his viewers a glimpse of the modern home of 2001. He predicted there would be an office where one could work from home. Such offices would contain multiple computers that could receive news and monitor the weather or stock market. There would still be a landline phone, but you could use it to make video calls. These predictions are all incredibly accurate, if a bit premature.

But the details were still quaintly off. Each function listed above had its own dedicated monitor, which you would control by turning knobs. The news reader could print a hard copy if you wanted one. The office, of course, was for the man of the house, who could also monitor other rooms in the house through closed-circuit TV, in which we see mother and daughter making a bed.

Beyond the occasional success, there is a lot to learn from these attempts at futurism. Understanding the many challenges to successful futurism can teach us about ourselves, our place in history, and how our present can shape our future. We now look back at the future predictions of the past as a window into their psychology, culture, and relationship with technology.

Hopefully we will get more right than wrong as we try our own hands at predicting the future of science and technology. At the very least this work will become part of the time capsule of futurism. Perhaps future futurists will look back at us and learn a little something about our present.

FUTURE FICTION: 2063 CE

Aayansh was not a fan of Mondays despite having a restful three-day weekend prior. He usually worked from home on Mondays, but not this one. At seventy-two, today was going to be the capstone of his career—the ITER fusion reactor was finally being connected to the Continental Synchronous Grid.

As he settled into a drone car for the final leg of his trip to Saint-Paul-lès-Durance, he checked his forearm, which displayed the time, local weather, and his itinerary for the day. He had eleven unread messages, sorted by category and urgency, but those could be dealt with later.

He almost didn't notice when they became airborne, the electric hum of the engines fading into the background, as he again looked through his speech. The ITER project had more lives than a cat, having survived endless delays and increasing calls for the expensive project to be shut down. There would likely be protestors there today. Europe's energy infrastructure was now 73 percent wind and solar, and purists wanted to divert all energy investment to getting that figure up to 100 percent despite adequate grid storage still being decades away. Meanwhile, the first-gen IV fission plants were getting to the end of their life span. Over the next few decades, they would either be shut down, extended, or replaced.

That is where the ITER came in—to keep those fission plants open or replace them with fusion. That was the argument that brought the ITER over the finish line. In another generation it would have been too late. Solar and battery technology were just getting too cheap.

Another message beeped for his attention, from Sarah at

Lunar-ITER, who was at the base overlooking surveillance for the proposed lunar reactor. She had been his number two Earth-side until she was assigned by the joint NASA-ESA board to head the ITER project near the Marius Station. The recorded message was in VR, so he put on his glasses, said "play," and was suddenly sitting on board the Marius station facing Sarah.

"I just wanted to wish you luck on your big day. Hope everything goes well down there, or our little project won't go very far." She pointed to the large window behind her showing the daylight lunar surface. "There it is. The regocrete 3D printer arrived last month. Once the plans are approved and everything is tested, we can start building the containment facility. Hopefully in twenty years I'll be cutting my own ribbon."

She signed off with a smile and a wave, and then Aayansh was visually back aboard his drone car. While his glasses were still on, he switched to augmented-reality mode and overlayed a map onto the landscape below. He could see the ITER in the distance, the sprawling facility, buried power lines connecting to the grid, nearby towns, the maglev line, and roadways. Just beyond the facility stood the enhanced outline of the dozen or so thirty-story hydroponic gardens that would ultimately be powered by the reactor. As he sat back, his perspective shifted and highlighted dozens of drone cars converging on the same location.

News drones flanked his car and he waved almost absentmindedly. Getting some privacy back would be a nice change.

He removed the glasses as the drone car took itself into an open landing pad. Today would be long and tedious, but then his Mondays would be free.

PART TWO

Today's Technology That Will Shape Tomorrow (and Tomorrow and Tomorrow...)

The future cannot be predicted, but futures can be invented.

—Dennis Gabor, Nobel Prize–winning physicist

We are inventing the future, right now. In this section, we will review the existing cutting-edge technologies that are most likely to shape our future. We will also review the history of each technology, like following the arc of a thrown object to see where it is headed. This process of inventing the future is never-ending, with multiple feedback loops interacting in complex ways. As we trace these trajectories, larger patterns do emerge, and they can partially illuminate the technology of the future.

At the very least we'll learn about some cool technology that is shaping our present.

4. Genetic Manipulation

We are still in the horse-and-buggy phase of genetic modification.

In 1984, the US government adopted plans for the Human Genome Project, joining what is still the largest international cooperative biological project ever. The goal was to completely map the genes that make up the human genome—the DNA that serves as a set of instructions for the development of a human being from a single cell. The project formally launched in 1990 with a goal of completing the mapping by 2005—they finished two years ahead of schedule in April 2003. They also finished under budget, although it still cost roughly $1 billion.

The technology for sequencing DNA had advanced considerably during the project and has continued to do so in the two decades since. Scientists have now sequenced the genomes of hundreds of plants and animals. In fact, we can now sequence a human genome in two days and at a cost of only $3,000 to $5,000. That's 2,372 times faster, and 250,000 times cheaper than in the early 2000s.

It's so fast and cheap now that we can do an exome mapping on individuals just to go hunting for any genetic diseases or anomalies. The exome is the important bit of the genes (the exons) that produce proteins, ignoring all the junk and regulatory DNA. For a few hundred dollars you can also have your own genetic markers mapped to determine your ancestry (although I can't vouch for the accuracy of any commercial lab).

This represents geometric progress, not merely linear or incremental, and has created a great deal of both fear and optimism about the future of genetics and genetic engineering.

Further, scientists are working on mapping the proteome—a complete list of all the proteins that make up humans or other species. Genes and proteins are the fundamental building blocks of life, and the ability to map and manipulate them essentially gives us the keys to biology. It is safe to predict, barring some extreme cultural backlash, that genetic technology, where it is and where it is going, will massively shape our future.

Genetic Modification

Humans have been genetically modifying the plants and animals important to our survival for thousands of years. This is part of the way we use technology to adapt the environment to ourselves much more quickly than adapting biologically to the environment. Traditional methods of genetic alteration include cultivation and interbreeding. By simply selecting plants with desirable traits to plant for the next generation, we can partially craft them to be better crops.

In fact, almost everything you eat has been significantly genetically modified in this way. Most modern crops look almost nothing like their wild ancestors. In the last century we have learned techniques that allow us to speed up this process. Perhaps the most important is hybridization—crossing related plants to produce a favorable combination of traits. Meyer lemons, for example, are a cross between traditional lemons and mandarin oranges. Most of the sweet corn varieties grown in the United States are hybrids of different corn cultivars.

Since 1930, farmers have also been using something called "mutation breeding." This technique employs chemicals or radiation to speed up mutations so that farmers can choose the tiny percentage with favorable traits to develop a new cultivar. Between 1930 and 2014, there have been 3,200 mutagenic species introduced into our agricultural system, including cultivars of wheat, pears, peanuts, grapefruit, and many others.

Toward the end of the twentieth century, we added further techniques to speed the process of altering the plant and animal varieties we depend on for food (and other things like medicine and fuel). This includes the ability to directly change the genes of an organism, by turning off a gene, altering a gene, or even inserting an entirely new gene from a related (cisgenic) or even distant (transgenic) species. These techniques have created pest-resistant corn, nonbrowning apples, fungus-resistant chestnut trees, and many others. It is these latter techniques that are generally considered "genetic modification" even though this is somewhat of an arbitrary category.

The resulting cultivars, often referred to as "genetically modified organisms" (GMOs), are a good example of the interaction between culture and technology. On the one hand, it is possible to discuss what this technology is currently capable of and will very likely be capable of in the future. On the other hand, we must examine how society has reacted to this technology, and how that will affect its future.

There are currently several technologies for making precise alterations to the genome of an organism. None of them are perfect, but the technology has been progressing very quickly. It begins with the first production of recombinant DNA in 1972. This technology combines the DNA from two or more species into one DNA strand, using enzymes involved in normal DNA regulation, and then inserts that DNA into a host cell (usually a bacterium) in order to alter its function. This technology revolutionized the production of some drugs that were otherwise difficult to manufacture.

Perhaps the most important early application of recombinant DNA technology was the production of insulin. Previously insulin had to be harvested from the pancreases of animals, and was slow and expensive to create, limiting its availability for the treatment of people with diabetes. With recombinant DNA technology, however, scientists could make a vat of baker's yeast inserted with genes that produce human insulin. This ability to produce large quantities of human insulin transformed the treatment of diabetes. A futuristic technology had come to pass, and it is still an amazing achievement of science.

Once the yeast is made, it is self-reproducing, but initially creating these genetic changes is difficult, expensive, and time-consuming, meaning that only very well-funded and equipped labs or large corporations can do it. We needed a programmable platform that could target specific sequences of DNA for manipulation. Three such platforms have now been discovered and implemented, starting with zinc finger nucleases in 1985. Zinc finger enzymes can recognize and therefore target a specific string of DNA, either in a test tube or a living organism, and deliver to that target enzymes that will splice the DNA, allowing for the deletion or insertion of desired bits of genetic code.

Zinc finger nucleases were a huge advance, but developing a genetic target could still take months and at a fairly high cost. Then, in 2011, a new faster and cheaper method was discovered: TALEN—transcription activator-like effector nucleases. Soon after, researchers introduced another gene-editing system that bacteria use as part of their immune defense: CRISPR—clustered regularly interspersed short palindromic repeats (which has been much in the news and is probably the one you have heard about already).

CRISPR is essentially a method for targeting where to go in a DNA chain and can be used to deliver a payload, such as Cas-9, which is an enzyme that can cleave the DNA. Other types of payloads can also be delivered. TALEN and CRISPR are both (relatively) cheap and fast methods for programming precise cuts to the DNA. If you think of DNA as a library of genetic data, these techniques are like a catalogue that will take you to the right book, chapter, and sentence you are interested in. You can then make edits to that sentence, remove it, or replace it. CRISPR is faster and cheaper, but TALEN may prove more precise in some uses.

With them, the time it takes to target a specific DNA sequence has decreased from months to days and is now within reach of most research labs around the world.

We are still on the steep part of the learning curve with these genetic modification technologies, and they are advancing so quickly that not only is progress geometric, but the pace is also accelerating. They are not yet perfect. There are, for example, problems with off-target changes, which means that CRISPR sometimes will target the wrong part of the DNA molecule in addition to the desired target (that library catalogue may have identified a dozen similar sentences in a dozen different books and targeted them all). But scientists are already learning how to tweak the trade-off between speed and precision with these systems. By altering the enzymes used and the structure of the CRISPR, they can slow down the process and reduce the number of errors.

Researchers are also developing new payloads that can modify genes in other ways. For example, researchers have created what they call CRISPR-On and CRISPR-Off. These techniques do not alter the target genes; they just turn off their transcription (making a protein from the gene). Now, we can reversibly silence a gene, turning it off and then back on at will. In some genetic diseases, like Huntington's disease, a disease of the brain that causes dementia and movement disorders, the biological harm is not only caused by losing the normal function of the protein, but the mutant protein itself may also be toxic. Turning off the mutant gene can therefore decrease the damage.

These technologies are already useful, and it seems inevitable that they will further improve, at least incrementally. What are the implications of having cheap, rapid, and powerful tools of genetic manipulation, in the short and medium term?

First, we must consider the impact on research itself. These are first and foremost tools of genetic research. They are already accelerating advancement in that field, which will continue to advance our understanding and therefore our ability to use genetic modification as a tool. For example, it is easy to see what function a protein serves if you can turn its gene on and off and look at the results. We not only need the tools of genetic change, but we also need to know what genes do, and how modifying them can alter health and function.

In the short-term, it is safe to say we will have faster development of genetically modified organisms. We already have GMOs that resist pests, are drought resistant, are herbicide tolerant, cold tolerant, and have a longer shelf life. Most of these alterations were targeted to farmers, making their farms more productive and profitable. However, many other possible applications are being developed.

One exciting category of GMO is nutritional fortification, such as Golden Rice. This is a rice variety with an inserted gene to produce beta-carotene, a precursor of vitamin A. This is meant to fortify a staple crop with vitamin A as one strategy to address the massive problem with vitamin A deficiency in the developing world. Golden Rice varieties are already available and going through the regulatory process.

In Africa, banana varieties are a staple crop—these are more like plantains, which are starchy, and not the Cavendish dessert banana with which we are all familiar. Staple bananas can represent as much as 40 percent of calories in some parts of the world. They are being threatened (as is the Cavendish) by fungal pests and will eventually be wiped out in their current form, but there are efforts underway to develop GMO bananas that have resistant genes inserted.

GMO papaya have already saved the Hawaiian papaya industry from the ring spot virus. There are also hopes of saving the Florida orange crop from citrus greening, another fungus.

The big picture here is that as we try to feed billions of people, we need to grow more and more massive amounts of food. Most of our crops, however, have already been cultivated over centuries or millennia to have weakened defenses. This is because plants make poisons to protect themselves. These poisons are bitter due to the ways animals have evolved the ability to taste poisons for their own defense. So, humans bred plants that were less bitter and safer to eat, but this deprived them of their natural deterrents. We then grow these plants in vast fields of monocrops of genetically similar plants, which is essentially an invitation to pests to come eat them. Suffice to say, this system is not sustainable.

There is a vigorous debate about how best to deal with this problem, and the results will shape our future. We will likely need to use many methods, including integrated pest management, increasing genetic diversity among our crops, use of cover crops and crop rotation, and use of pesticides. One solution would be to genetically engineer crops that are resistant to pests while still being safe and tasty for humans to eat.

This does not mean we need to use the latest and most advanced technology for the sake of progress. Sometimes traditional breeding techniques or hybridization are still the best methods or will do the trick easier and cheaper. But when entire industries of staple crops are threatened, we will also need to turn to GMO technology, as we already have. Over the next forty to fifty years, we will very likely see said GMO bananas, oranges, and others introduced in order to save these crops—that is, unless political opposition becomes too fierce.

It seems unlikely, though, that populations will be allowed to starve for a principle. I suspect the outcome will be similar to what happened after the first in vitro fertilization (IVF). There was much opposition to the first "test tube" baby, which was decried as unnatural, with fears of a population explosion. But the technology quickly became commonplace. None of the dire predictions came true. All that happened is one more option for couples who could not conceive naturally.

In the same way, GMO technology is continuing to advance in

the background, and it will be very hard to deny the world the benefits of what it produces. Even one of the most politically anti-GMO states on the planet, Hawaii, quietly accepted the GMO papaya rather than see one of their most important crops die.

There are also active efforts underway to develop GMO crops that use more productive forms of photosynthesis. This could produce a 20 percent or so increase in productivity for the same amount of land. Yet another project is developing to transfer the ability to fix nitrogen from the atmosphere by combining it with compounds to make it available for chemical reactions. Only some plants acquire nitrogen through symbiotic bacteria in the roots, but if our major staple crops could get all the nitrogen they need from the air (which is 78 percent nitrogen), that would dramatically reduce the need for nitrogen fertilizer and would further increase productivity.

GMOs also serve functions other than providing food. Genetically modified bacteria and yeast already are used to manufacture drugs. Researchers are working on bacteria genetically modified to eat oil in order to clean up environmental spills. They could also eat other toxins or be programmed to glow in the presence of such toxins, alerting us to their presence. GMOs can consume garbage as well and create biofuels from the waste while they're at it.

While I don't think we will be seeing giant crops in the future, as in the movie *Sleeper*, it is very likely that much of our crops will be altered to be pest resistant, have more efficient photosynthesis, fix their own nitrogen, have a longer shelf life, improve flavor, and perhaps have enhanced nutrition. At the same time, we will still be traditionally breeding, planting heirloom crops, and using other conventional farming techniques.

I also don't think our agricultural woes are disappearing anytime soon. Producing enough food with the lightest footprint on the environment, while keeping one step ahead of pests that want to eat our food, is likely going to be an endless struggle.

Treating Disease

In 2018, He Jiankui of the Southern University of Science and Technology in Shenzhen, China, became the first scientist to use CRISPR to treat a human disease. He altered the DNA of twin girls using CRISPR as part of in vitro fertilization. Their father was infected with HIV and the gene alteration was meant to decrease the probability of transmission.

His actions were widely condemned. It was premature to apply this new technology, which was not reviewed or approved (representing unethical research), to humans and there are already other established methods of preventing HIV transmission in donated sperm for IVF. But his decision shows the allure of using advanced genetic modification techniques to treat human disease.

Of course, the most obvious target for such treatment is genetic diseases themselves. Scientists have been trying to perfect gene therapy since the 1990s. Some early attempts used a retrovirus to insert DNA into the patient to correct for the genetic defect. One example is cystic fibrosis, which results from a mutation causing thick mucus and mainly affecting the lungs and resulting in early death. Unfortunately, some early clinical trials to treat this disease resulted in deadly viral infections, sidelining the technology for decades while these kinks were worked out.

Gene therapy requires a vector, like a virus, to get the genetic modification technology to the target cells. In other words, CRISPR can target the correct stretch of DNA, but you have to get the CRISPR-Cas9 proteins into the correct cells where you want to make the change.

One method is to perform genetic modification on a single cell in a petri dish, as Dr. He Jiankui did. If you change the embryo before it is implanted, you could theoretically rewrite every cell in the resulting organism. You can also extract cells from either the blood or bone marrow, do the genetic modification, then put the cells back in the body.

Performing modification on cells in a living organism is trickier. You must get the modification platform into the body and get it to the cells you want to change. Viral vectors are the most common. Viruses evolved over millions of years to get into specific cells in their hosts, so they are good at it. There is a lot of interest in nonviral vectors, such as using RNA itself, because they are safer and can deliver larger payloads. Yet this is still very much a nascent technology.

Obviously, there are also safety issues beyond introducing viruses. The genetic modification platform itself has off-target effects, meaning it alters genes that are not the intended target. Additionally, whatever vector we use might deliver the payload to cells we don't want to target.

Given the advances so far in this technology, and the speed of these strides, it seems reasonable that these technological hurdles will be overcome. Even incremental advances will be enough to make gene therapy a common reality. In fact, it is already happening. Several individuals have been successfully treated for the genetic disease sickle cell anemia with CRISPR.

Over the next twenty to fifty years, it is therefore very likely that one form of gene therapy or another will be used to tackle the more common genetic diseases one by one. These won't always be outright cures. It is more likely that genetic changes will be partial, but enough to dramatically reduce the severity of symptoms. Genetic changes can also be introduced that don't fix the mutation but compensate for its effects. For example, we could alter astrocyte support cells to help keep diseased neurons functioning longer, slowing the progression of neurodegenerative diseases like Alzheimer's.

Combined with IVF, genetic modification can also be used to entirely prevent some genetic diseases before implantation. This would involve a technique similar to what He used but would swap in a healthy gene for one carrying a mutation for a horrible disease. This will not end genetic disease because many are recessive and therefore undetected until a random alignment from both parents. There are also new spontaneous mutations, sometimes occurring during development, so after IVF would have taken place. Still, genetic modification could dramatically reduce the burden of genetic diseases and render them manageable.

It's hard to deprive a child of an effective treatment or even cure for a horrible disease, so I don't anticipate any effective pushback against these applications (assuming they follow proper ethical and scientific protocol). The question is, how far will we take this technology?

There are other medical applications aside from genetic diseases. Already there is research looking at the use of CRISPR to treat cancer. CRISPR can target specific DNA, which can include mutations that cause cells to become cancerous. All they have to do is target these cells and splice the DNA enough to kill the cell. Early results with this application are very promising.

I will never predict a "cure for cancer" because cancer is not one disease but many, and it has proven very difficult to treat. However, survival from most cancers has been steadily increasing as each new treatment adds to our tools and pushes back cancer one bit more. Genetic tools will likely be the same, but as they are proving to be very powerful, I am hopeful that their effect may turn out to be more than incremental.

Beyond genetic diseases, where a specific mutation causes a specific disease, there is genetic predisposition. Some people, for example, are more likely to have a heart attack than others. This is the reason doctors take a family history in order to estimate risk for common diseases. Reducing the risk of these common afflictions poses another opportunity for medical genetic modification. Imagine if we could reduce instances of diabetes by half, or high cholesterol, hypertension, or the risk of Alzheimer's disease, all of which have powerful genetic components that could potentially be altered.

Fancy genetic modification would not replace having a healthy lifestyle. Eating well, exercising regularly, getting good sleep, and avoiding things like smoking and alcohol are always going to have a profound effect on our health. Genetic therapy could work in tandem, making us more resilient and perhaps even helping us achieve these lifestyle goals.

Remember the basic futurism principle: New technologies do not always replace older technologies but exist alongside them. Various types of gene therapies are very likely to become a powerful tool for preventing and treating disease in the future, along with existing established methods, and of course other future developments.

Genetic Enhancement

It seems like a very high probability prediction that we will use genetic modification to increasingly treat and prevent diseases as well as to enhance crops and even animals. It is more difficult to predict, however, to what extent we will use genetic modification to enhance humans. This is where I anticipate the most ethical push-back, and already there are laws prohibiting such applications. In the United States, for example, genetic modification in humans that can be inherited are banned outright. Specific applications of human gene modification have to be individually approved, and the FDA approves only those designed to treat a disease.

There can be a fine line between disease prevention or even treatment and genetic enhancement. This is partly because there is a fuzzy line between simple genetic diversity and the difference between healthy and unhealthy. How short must someone be, for example, before we consider their short stature a disability? For some conditions that most people would consider a disorder, like genetic deafness, Down syndrome, and even autism, there are those who consider these to be part of normal human variation, just atypical. There is already pushback by some in the deaf community against erasing deaf culture by curing deafness.

Therefore, in modifying genes to alter a trait, are we treating a disorder or making an enhanced human? If the change is considered a lateral move, then it might even be considered a "designer" human, with traits chosen for personal choice, cultural preference, bigotry, or even fashion. For some features, like eye color, the choice is clearly aesthetic. For others, like height, build, and strength, the lines between treatment and enhancement are blurred.

From a technological point of view, we are on a path to at least steady incremental improvements in our ability to alter the genome to produce a desired outcome. If we continue to develop this technology, there is little question that in fifty years or a century we will have a profound ability to genetically design people. The ultimate, mature form of this technology will probably be nothing less than near complete control over what are essentially programmable people.

The real question in terms of predicting the future, therefore, is how this technology will be accepted and regulated. Where will we draw the line, and will attempts at regulating genetic modification be successful or simply spawn a black market?

It's clear that treating disease and even preventing certain diseases will be generally accepted. There will always be technophobes, but they will likely be on the fringe when it comes to gene therapy. It's even possible that genetic modification to reduce disease will be strongly encouraged to reduce health care costs, even required in more authoritarian countries. Basic enhancements, however, will be squarely in the controversial gray zone. There will be fears of creating "super soldiers" or that the rich and powerful will become genetically elite, as in the movie *Gattaca*. Probably the most controversial will be enhanced intelligence. We tend to be more accepting of elite athletes or elite artists, but very suspicious of elite intellectuals.

Neutral but aesthetic choices, like eye or hair color, will probably involve less heated debate because the changes are ultimately neutral. But they will likely still generate cultural debate. Already, for example, it is popular but also controversial in some parts of Asia for young people to have cosmetic surgery to make their eyes look more Caucasian and less Asian—so-called double eyelid surgery to create that crease in the eyelid that many Asians lack. Some feel this represents caving to a biased Western cultural standard of beauty.

Imagine if you could simply tweak your child's genes to give them all the culturally trendy features. This will likely (and probably correctly) be seen by some as cultural hegemony, even cultural erasure. Popular countries might not just export their culture, but their genes. Also think of the controversy it would generate, for example, if it became common among African Americans to have their children genetically modified to pass as white to evade persistent racism. Caucasian couples might, on the other hand, want more diverse children. Would this be the ultimate cultural appropriation?

In the end, might we end up like Dr. Seuss's Sneetches? Genetic modification might become so common that cultural heritage is no longer discernible (we may entirely forget who has stars upon thars). Will this be a good thing or a bad thing? Probably a little of both.

The big theory here is that human genetic modification, if allowed to be widely or freely adopted, would disconnect our genetics from our heritage. We will no longer be trapped in the genes we happen to inherit from our parents. This would scramble the current debates around race, culture, and history.

If taken to an extreme, human genetic modification might also challenge the definition of what it means to be human. Would humans be able or allowed to incorporate genes from other species? What about new, entirely crafted genes? Will there be humans with tails, fangs, wings, scales, and other decidedly nonhuman traits?

There might also be nonhuman super-enhancements. I am not talking about X-Men type stuff with laser beams coming out of our eyes, but rather biological traits that would be considered superhuman. It is plausible, for example, to genetically engineer tissue so that it can regenerate. We could redesign human anatomy from the top down, creating humans with two hearts, redundant arteries, reinforced ligaments, and tendons, built-in cushioning for the brain, super-livers that render us immune to most poisons, or augmented immune systems to ward off infections and cancer. The list of probable upgrades is long.

We can also use genetic modifications to adapt humans to other environments, such as living on the moon or aboard space stations, or even in desert or frigid environments.

We might also genctically modify not just ourselves but also the bacteria that live in and on us. Our symbiotic flora could be engineered to eliminate tooth decay, give us all sweet-smelling breath or flatulence, better digest food while maintaining an optimal weight, further help to fight off infections, and secrete all sorts of helpful biochemicals.

There is another extreme aspect to the potential for genetic modification of humans—we might modify the structure of our DNA, not just individual genes. For example, much of our DNA is "junk" DNA because it does not code for proteins. Some estimate as much as 98.5 percent. This figure is controversial because we are not sure how much of the DNA is regulatory or serves some function other than direct coding, but there is good evidence that at least 75 percent of our DNA is junk, and the real figure is likely closer to 90 percent.

What if we engineer human genetics to eliminate all the clear junk, leaving behind only DNA that we need? This would dramatically reduce the biological cost and complexity of cellular reproduction. Our cells wouldn't need to waste resources copying the unnecessary portions and would reduce the number of copy errors in replication.

There are plenty of unknowns in what I am suggesting, but it is highly plausible that an optimized and vastly reduced human genome is possible. If created, this would likely be incompatible with natural human DNA—so people with the new optimized genome would not be able to have fertile children (or perhaps any children) with people with a natural genome.

Such a modified genome would, either by design or as a side effect, create a separate human species that cannot interbreed with the existing human species. It would create the ultimate in sci-fi elitism: the Eloi and the Morlocks.

The Uplift Wars

We might turn our genetic mastery to animals in addition to humans. I've discussed the application of genetic modification to agriculture, chemical manufacturing, and environmental cleanup. I left until now perhaps the ultimate in genetically modifying animals: making them as or more intelligent than humans.

In a series of novels, author David Brin explores the idea of "uplifting" animals into full human sapience. In fact, in his fictional universe, most intelligent species were uplifted by older species (their patrons) in an endless chain of uplifts going back as far as they have records. When human civilization first encountered this one interwoven intergalactic civilization, we were considered an aberration, "wolflings" who were abandoned by our patrons. The other species would not accept the possibility that we naturally evolved our intelligence.

By the timeline of the book series, humans had already uplifted chimps and dolphins and were well on the path to uplifting gorillas. While currently science fiction, this is not implausible. There are many species that have high neuronal density and might not take that much genetic manipulation to have human-level intelligence—dogs, racoons, bears, and most primates, to name a few.

There is no scientific or technological reason why uplifting species into human-level intelligence would not be possible. It may prove more or less difficult, and therefore it is the timeline of this advancement that may be hard to predict. But eventually, if this is something we want to do, we could be living in a planet of the apes or have snarky racoons flying our spaceships.

Obviously, the potential here is immense, and there are so many hypothetical outcomes it becomes impossible to predict. The ethical implications are also horrifically complicated. There are concerns about doing this in the first place, and if we did, what would our responsibility be toward the creatures we uplift? Would they be automatically granted human rights? How would they be integrated into human society, or would we give them their own space?

As we try to imagine a future with advanced genetic engineering, the limits on futurism become clear. In the near and medium term, however, I think we can make some high-probability extrapolations: Genetic modification will continue to become more powerful and accessible, leading to increasing genetically based medical treatments. Despite resistance, GMO crops and even animals will be increasingly incorporated as one of many solutions to our agricultural challenges, in addition to drug and other manufacturing and environmental applications. And we will at least begin to explore the possibility of improving the human condition through some limited forms of genetic enhancement.

The long-term, however, is a blur. Much will depend on how culture and individual morality evolve. Perhaps someday being narrow in our definition of what it means to be human will be looked on as quaint and backward. It is equally possible that there will be a reaction to the potential of genetic manipulation, with large portions of the population embracing our natural state and decrying any attempts at "messing with nature." I'm sure the term "Frankenstein" will play a prominent role in whatever future debates are had on the topic.

Likely everything will happen and there will be every kind of reaction. The real question is one of proportions. Will these extreme outcomes be outliers or mainstream? Will there be isolated "Islands of Dr. Moreau," or will uplifted chimps be walking next to us on the sidewalk, hardly even worth a notice?

There is, of course, one application of genetic modification that I did not touch upon because it deserves a chapter of its own—modifying cells to become stem cells that can then be used in their own medical applications. In the next chapter, we will explore stem cell technology, the promises yet to be kept, and where it might all be headed.

5. Stem Cell Technology

Amazing potential, but trickier than we thought.

In 1958, French oncologist Georges Mathé performed the first stem cell therapy. Perhaps more surprising than how far back stem cell therapy goes is the fact that over half a century later, at the time of the publication of this book, the type of therapy Mathé used remains the only proven stem cell treatment. Therefore, despite being an icon of futuristic medicine, stem cell therapy remains 1950s technology.

Stem cells are cells all creatures naturally possess with the potential to develop into different cell types, which is how any tissue healing or regeneration occurs. Adult stem cells likely occur in all tissues but have been clearly demonstrated in bone marrow, peripheral blood, the brain, spinal cord, dental pulp, blood vessels, skeletal muscle, epithelia of the skin and digestive system, cornea, retina, liver, and pancreas. Some adult stem cells may also be able to generate cells that can differentiate into a few different types of mature cells (multipotent). Adult stem cells that can form only one cell type are unipotent.

A multipotent stem cell can develop into several different related cell types. For example, a bone marrow stem cell can turn into different types of blood cells, but not other cell types. A pluripotent stem cell can turn into any type of cell in the body. A totipotent stem cell, which only comes from embryos (and therefore are also called "embryonic stem cells"), can develop into any cell type in not only the fetus but also the placenta.

Stem cells also have another very interesting property—they are immortal. They would have to be, when you think about it, because multicellular life has a continuous line of inheritance, through embryonic stem cells, going back at least 600 million years. They are, by definition, cells that can survive for the life of the organism and can make clones of themselves to replenish the tissue in which they are embedded. In culture, stem cell lines can also be maintained indefinitely.

Back to the 1950s, when Mathé was doing research on bone marrow transplantation in animals: Bone marrow stem cells, called "hematopoietic stem cells"—blood producing—constantly replenish the supply of red and white blood cells. Mathé found that if he transplanted bone marrow from one animal to another of the same species, which is called an "allogeneic transplant," the immune system of the recipient animal would reject the transplant. However, if he first destroyed the bone marrow of the recipient with radiation, they would be able to accept the allogeneic bone marrow transplant.

The reason radiation pretreatment is effective is because it wipes out the immune system (the white blood cells) of the host, who would then essentially receive the immune system of the donor. Wiping out the host's immune system with radiation is very dangerous, however, and so the procedure was not performed in humans.

However, in 1958, an opportunity arrived. A reactor accident exposed several Yugoslav physicists to radiation, wiping out their bone marrow. Mathé performed an allogeneic bone marrow transplant into the survivors. The bone marrow survived and to some extent started producing blood cells, a process called "engraftment," and therefore became the first successful stem cell transplant in humans. The transplanted immune systems did start to attack the host bodies, however, and so Mathé was also the first to document what is known as "graft-versus-host disease." This basic procedure, radiation followed by allogeneic bone marrow transplant, is still used today to treat certain types of blood-borne cancer.

Why, then, is there so much hype surrounding stem cell therapy if we are still limited to using a treatment developed in the 1950s? In the interest of transparency, there are also a few types of tissue transplants, like skin and retinal, that require stem cells in the transplanted tissue to heal and function. These are a limited type of indirect stem cell transplants, but not really considered stem cell therapy.

The increase in interest in stem cells partly goes back to our advancements in understanding genetics and genetic modification. The basic science of what makes a stem cell multipotent or pluripotent has been steadily progressing and was given a boost by the Human Genome Project. To further this research, scientists were relying on embryonic stem cells, which have the greatest stem cell potential. However, conservatives were concerned about harvesting cells from aborted fetuses or from fertilized embryos that were not implanted. When George W. Bush was elected in 2000, he quickly took up this culture war. With a new Republican president, the time seemed right to push back against stem cell research.

This, in turn, motivated some scientists and others to argue for the fantastic potential of stem cell therapy. It could hypothetically cure degenerative diseases like Alzheimer's dementia, lead to the ability to regrow failing organs, and usher in a new era of modern medicine. Banning this research would just cede this critical technology to our competitors.

This is when stem cell therapy entered the public consciousness.

On August 9, 2001, Bush signed an executive order that banned the creation of any new cell lines from embryonic stem cells but allowed research to continue on cell lines that were already in existence. This was considered a "splitting the difference" solution, but the net effect was to significantly decrease federal funding for stem cell research in the United States. Some states, most notably California, took up the slack with their own funding, and the order was ultimately reversed in 2009 by President Barack Obama.

In the meantime, however, scientific progress mostly (but not entirely) rendered the question moot. During this time, the technology of adult stem cells was advancing considerably. The turning point came in 2006 when researchers Shinya Yamanaka and Kazutoshi Takahashi were able to turn an adult multipotent stem cell into a pluripotent stem cell in mice. A year later this was accomplished with human adult stem cells. This new type of stem cell was called "induced pluripotent stem cells" (iPSC). Amazingly, this feat can be accomplished with changes to only four genes.

The development of iPSCs was a game-changer for a couple of big reasons. The first is that it mostly obviates the need to harvest PSCs from embryos. PSCs still have some advantages over iPSCs for research, but the entire controversy can be avoided with iPSCs.

Perhaps more significant, however, is that iPSC technology allows for the creation of pluripotent stem cells from an adult. As demonstrated by bone marrow transplantation, rejection is a significant issue. If, however, it was possible to transplant stem cells derived from the host's own cells, then they would be immunologically identical—there would be no issue of rejection.

Breakthroughs in the basic technology of stem cells only served to magnify the hype about their medical potential. A significant downside to this was the popping up of numerous bogus stem cell clinics, usually in countries with poor medical regulations. Desperate people suffering from serious illnesses continue to be lured to these clinics with the promise of fabulous stem cell treatment, often at a cost of tens or even hundreds of thousands of dollars. Unfortunately, there is no evidence that any of these treatments are safe and effective, and in most cases, what they are injecting cannot even be confirmed.

In fact, these predatory clinics are taking advantage of one of the most common futurist fallacies: the tendency to overestimate short-term progress. It often takes twenty to thirty years for basic medical research to reach the clinic, making these clinics decades ahead of schedule.

Not only does it take time to work out the details and prove safety and effectiveness for each new medical treatment, but there are also significant technological hurdles that remain for stem cells. More pessimistic experts are even concerned that stem cells may never overcome these hurdles and reach the potential that has been promised.

The first big obstacle relates to the reason that evolution did not see fit to make us flush with stem cells in the first place, with virtually unlimited potential to regenerate injury. Why can't I regrow an arm like some lizards can regrow a lost tail? This is because stem cells are a double-edged sword. Their immortality and ability to make copies of themselves without limit mirrors characteristics of another type of cell: cancer. The fact that stem cells survive for a long time also means there is more time to accumulate genetic mutations that will render them cancer cells, and these cancer-causing mutations have been found in pluripotent stem cells. Evolutionary forces are good at optimization, and so we likely have the ideal number of adult stem cells, enough to keep our tissues alive and healthy while minimizing the risk of cancer.

Injecting pluripotent stem cells into people carries a definite risk of cancer, and a lot of research is focused on identifying and mitigating this risk. While this research is progressing, the problem has not yet been solved.

Another issue with stem cell therapy involves getting stem cells to do what we want. For bone marrow, the stem cells need to just survive and reproduce, which is probably why this is the only proven therapy, but other cells need to do more. If, for example, we inject neural stem cells into a brain injured from a stroke, they need to go to the correct location in the brain and make meaningful connections with other brain cells.

Tissues that have complex structure to them therefore pose a technological challenge for stem cell regeneration. For this reason, the low-hanging fruit will likely be for tissues with the simplest structure, like skin. Heart muscle may also be plausible, because heart cells will connect to and spontaneously synchronize their beating with other heart cells. Adding a little bulk to an injured heart, therefore, may be one of the earlier stem cell applications.

Our basic science understanding of what makes a cell a stem cell, the underlying genetics, and how to manipulate them is advancing considerably. Clinical applications, however, are still years and perhaps decades away because this is a very tricky technology to use. Still, there are potential uses for stem cell therapy for us to explore and see where it may be headed in the near and long-term.

Injecting Stem Cells

One of the more obvious applications of stem cells is to inject a slurry of such cells directly into a person, either into the bloodstream or into a specific organ or tissue, in order to cure disease or injury or even reverse the effects of aging.

The idea here is simple, although for the reasons just stated may prove difficult to pull off. But if we can get it to work, the idea is that stem cells will differentiate into the appropriate cell type for injured tissue and then replace the damaged cells. This kind of therapy could bulk up wasted muscle, replace damaged heart cells after a heart attack, replace brain cells lost to a stroke, regrow skin damaged by burns, or replace the insulin-producing cells of the pancreas in a person with type 1 diabetes. You could potentially apply this principle to any organ or structure in the body.

Similarly, stem cells could replace diseased or abnormal cells. In the previous chapter, we covered curing genetic diseases through gene therapy, and the challenge of targeting all the cells that need to be changed. Alternatively, you could just reprogram a small number of stem cells taken from the patient and then reintroduce them into the body to replace enough of the diseased cells to fix whatever disease or deficit the mutation causes. This approach has already been used to treat sickle cell anemia, a genetic mutation that causes misshapen red blood cells that can clog capillaries.

That same idea can apply to an adrenal gland making too much or too little of some hormone. Patients whose lungs do not make enough surfactant, which is necessary to keep the air sacs open and functioning, could be given stem cells that do, or we could replace the muscle cells in a patient with muscular dystrophy.

Yet another approach does not involve replacing cells but just injecting support cells into diseased or damaged tissue. This would be an easier process in that the stem cells do not need to replace existing cells or make complex connections or anatomical structures. They just need to take up shop and survive. Such cells could be programmed to release hormones to support the function of other cells, change the local environment, or even secrete drugs. Utilizing stem cells as support cells is a promising approach.

Stem cell treatments could also be used to reverse some of the effects of aging, such as growing bone or replacing cartilage in worn-out joints. This type of treatment could also augment the function of organs that are not diseased, just old. Collagen could even be added to skin to make it look more youthful.

We may not even need to inject stem cells, if we can figure out how to stimulate the adult stem cells already in the various tissues to reproduce and replace any aging, injured, or diseased cells. Assuming we can do this without just growing cancers. The mature form of this technology could theoretically replace most of the cells in your body with youthful stem cells, extending life span considerably while maintaining vigor and health.

While not feasible in the short or even medium term, this is theoretically possible for the distant future. We do have an example from the animal kingdom that might inform the viability of this approach—the immortal jellyfish (technically a "jelly"). *Turritopsis dohrnii* is a small jelly that, when injured or diseased, reverts to an immature polyp stage and then regrows into an adult clone. This self-cloning can repeat indefinitely, as far as we can tell.

This type of stem cell regeneration, however, is more like the next category of stem cell therapy: growing entire organs.

Growing Organs

The World Health Organization estimates that over 100,000 solid organ transplants take place around the world each year. Organ failure can be fatal, or require extreme medical interventions to keep patients alive, like kidney dialysis.

While lifesaving, organ transplants can have significant limitations. The first of which is the availability of donor organs. Some are cadaveric, from the recently deceased, while others are live donors. In the United States alone, about 7,000 people die each year waiting for a transplant that did not become available in time. For those lucky enough to get one, they will need to spend the rest of their life on immunosuppressing drugs to keep their body from rejecting the transplant or the transplant from damaging their body.

Rather than injecting stem cells, we could use them to build or grow an organ outside the body for later transplant. This is where iPSCs are extremely useful. It is possible to take skin cells from a patient, turn them into induced pluripotent stem cells, and coax them into differentiating into the needed cell type. In this way we can grow an organ with a person's own cells, and therefore their immune properties. This would obviate the need for any immunosuppressive treatments, as there would be no rejection.

This, however, is where some additional technology is needed. We need more than just a clump of liver cells; we need a liver. Putting the stem cells into a dish will not allow them to grow into a whole organ. In normal development from an embryo to an adult, organs form in a specific location and respond to nearby anatomy, the local physiological environment and chemical signals. In other words, organs are programmed to develop inside whole organisms, not in a dish.

What are the options to make whole organs? One is to grow them attached to living animals. You may remember the image of a mouse with an ear growing on its back. The living host can provide nutrients and oxygen while the new organ grows. Some organs may need the environment inside an animal's body. This would likely require some genetic alteration of the host animal to keep them from rejecting the organ while it grows.

Another option is to use scaffolding—something to provide the structure of the mature organ while the stem cells grow on it. This is already being developed with studies using donated organs denuded of living cells while the connective tissue is left in place. That connective tissue is the scaffolding, resulting in a fully formed organ. We are not there yet, but this approach is showing promise.

Three-dimensional printing also can be and has been adopted to printing with cells. Stem cells can be layered on top of each other and "printed" into the structure of the desired organ. This has been combined with the scaffolding technique—printing cells onto the scaffolding.

Another theoretical option that has not yet been developed is to create artificial environments that allow for entire organs to grow. These would be like the Axlotl tanks of the Dune science fiction series that were used (allegedly) to grow replacement organs.

It is also possible to grow entire animals with human organs. This has more to do with genetic modification than stem cells, but I include it here as it is a method for possible organ transplants. For instance, researchers have been looking into genetically modifying pigs to grow hearts with human immune traits. Ideally these human-pig hearts would be compatible with humans in general and would be engineered not to provoke an immune response. Such animal organ donors could even be engineered to match the individual immune markers of the intended recipient.

In early 2022, the first living human recipient, David Bennett, was given a heart transplanted from a genetically modified animal donor—a pig that had ten genetic alterations. He still required powerful immune suppression to avoid rejection, and unfortunately he died in March 2022.

One final example of growing replacement organs was shown in the 2005 movie *The Island*. In this film, the character played by Ewan McGregor lives in a closed future society, where he learns the horrible truth that he is a clone of a wealthy donor living outside in the "real" world. All the people in his society are clones, who are kept alive and ignorant of their nature until their owner needs an organ. They are told they have "won" a lottery and will be sent to live on an idyllic island. However, they are just being sacrificed so that their organs can be harvested.

Is a more humane version of this method possible, one with fewer blatant ethical problems? The clone could be kept in a comatose state, never experiencing their own existence, but this would be difficult and expensive. Or they could be engineered to not have a brain, and therefore arguably not be a person but just meat. They could just be a living torso, complete with internal replacement organ, in a vat. Still, this idea is not likely to have wide support. It seems most probable that some combination of printing organs using some sort of scaffolding will find application in the near term (which may still be decades).

Growing human-compatible organs in animals like pigs is the most likely application to be technically possible in the nearest time frame, but this is limited by animal rights issues and the enduring queasiness about mixing animal and human parts (despite the fact that pig valves and other animal parts are routinely used in people). The controversy over the use of animal parts in humans began on October 26, 1984, when Stephanie Fae Beauclair—known as Baby Fae—was given a transplant of a baboon heart. It kept her alive for twenty-one days but ultimately failed. This controversy has contributed to the slow progress being made in this technology since then.

If history is a guide, however, such controversies will die down in direct proportion to this technology's success. The more people who are saved with animal-derived organ transplants, the less compelling theoretical or abstract objections will seem. I think queasiness over animal transplants will then go the way of hesitancy over in vitro fertilization.

The need for organ transplants is great, and the ability to do away with immunosuppression would also be profound. This is a branch point in the direction of future technology, as there are multiple ways to accomplish this goal. At present I would bet on growing genetically engineered human-compatible organs in animals as the most likely short-term winner, with tremendous long-term potential.

Immortality

Would the ultimate expression of combining genetic and stem cell technology lead to virtual immortality? Will *Homo sapiens* join the *Turritopsis dohrnii* jelly in being able to completely replace all the tissue in our body with essentially fresh cells? Could we grow ourselves an entirely new body?

The ultimate problem here is the human brain. Jellies don't need to worry about continuity with their past memories. If we just grew an entirely new body, including a new brain, it wouldn't be us but instead a clone. Ideally, then, our brain (which *is* us) would need to grow a new body for itself. Is this even feasible? Not with any extension of today's technology. But it is theoretically achievable.

One prospect is that we could grow a cloned body, then transplant our existing brain into the new body, although brain transplants are not currently possible and may be centuries away if ever. We could also transfer our memories into the brain of the clone. This is not really a solution, however. You cannot "move" yourself into another body. Again, you are your brain. At best you'd make a copy of yourself. Some might argue that this is a type of immortality. A version of you would endure, but there is no question that it would not be you. You would die; your clone would live on.

Another possibility is to grow a new body around our existing brain. Perhaps we could develop a form of stem cell regeneration in which we enter a chrysalis-type state where everything but our brains turns to goo, which feeds the stem cells. They would then grow an entire replacement body, including connections to the existing brain. Of course, it would take eighteen years to get a full adult body, unless we also had the technology to force growing at an accelerated rate. Again—we are talking far, far future here. This would require a level of control over biological processes that we can only imagine today.

Stem cell potential is truly transformative, and theoretically could result in a form of human immortality (although we'd have to balance that against going through puberty over and over). And yet the technology is so tricky it is difficult to predict when the payoffs will come, and how complete they will be.

Further, while genetic engineering and stem cell technology may allow our brains to live longer and healthier lives, they are not a pathway to neural immortality. For that, we would need a new approach, a digital one. We will need to interface our brains with computers.

6. Brain-Machine Interface

Our path to becoming cyborgs has already begun.

In the future, we will not only use technology, but we will also become technology. We will merge with our own machines—in fact, we have already begun to do so.

In 1960, two scientists from Rockland State Hospital in Orangeburg, New York, Manfred Clynes and Nathan Kline, published the paper "Cyborgs and Space," in which they address the challenge of people living in the harsh environment of space. The Austrian-born Clynes, a neuroscientist who was also a classical musician, and Kline, a psychiatric researcher from Philadelphia, argued that it would be "more logical" to adapt humans to space than to try to adapt space to humans. In order to do this, they asserted, we must make "man-machine systems," and they proposed the term "cyborg," a portmanteau of "cybernetic organism," to refer to these part-biology, part-machine entities. Cyborgs have been a staple of science fiction ever since.

The notion of cyborgs has deep implications for futurism. As we explore the possibilities of future technology, we cannot consider only what technologies will exist, but also how our relationship with technology will change. We previously discussed the ways in which we may manipulate the genetic code to alter ourselves, our pets, our food, and our environment. Now we will take a peek at how we can alter ourselves by merging with our nonbiological technology.

Our path to becoming cyborgs has already begun. Many people have cardiac pacemakers, deep-brain stimulators, prosthetic limbs, internal fixation rods to support a degenerated spine, and other implantable medical devices. Increasingly we are developing fully mechanical organs, such as artificial hearts.

Implantable devices are becoming smaller, more capable, with more portable power supplies. Engineers are also learning how to harvest small amounts of energy from the environment—from heat, or small movements—which will enable implantable devices to gather the energy they need from biological functions, getting one step closer to the "man-machine system" envisioned by Clynes and Kline.

To *fully* merge with our technology, for it to become a seamless part of us, we need a functional brain-machine interface (BMI). Our brains will need to directly control our machine parts and receive information from them. Just as the brain both feels and controls every part of your natural biological body, there would ideally be no difference for the mechanical parts.

In the past two decades, BMI technology has made tremendous progress and this success has painted an increasingly optimistic view of the future of cyborgs. We didn't always know if this was necessarily going to be the case. The critical question has always been will the human brain accept a mechanical interface, and will it feel real to the user? Having a robot arm strapped to your body is not the same thing as a robot arm being part of your body, being part of you.

Science fiction depictions of cyborgs have mostly ignored the question, assuming the interface would feel natural. When Luke Skywalker gets a new mechanical hand after Darth Vader (another cyborg) chops it off with a lightsaber, Luke clearly feels and controls his new droid hand as if it were a natural part of him. In this and other cyborg depictions in fiction, there is usually no indication of brain-machine interface. But this is the most critical part of the technology.

What needs to be in place for the BMI to function? First, brain impulses would need to be able to activate electronics, whether it is part of a computer or a robotic prosthesis. This is the easy part—brain impulses are electrical, and you can measure the brain's electrical activity through scalp electrodes. We have been doing this since German psychiatrist Hans Berger first did so in 1924.

There are, in fact, several ways to potentially control a machine with the brain. We can record the brain waves directly with electrodes and somehow translate those signals into the desired action. We could also connect machines to nerve endings that provide the electrical signal. Or, where practical, the electrical impulses generated by muscle contractions can provide the control signal. All technology we will explore later.

The reality is that the body is essentially already an electrical machine. Every cell has an electrical potential across its outer membrane. Skeletal and heart muscle, nerves, and brain cells evolved to use that electrical potential to conduct information, and that adapts well to our electronic technology, whether analog or digital.

How about getting information from machines to the brain? This can be a bit trickier, as there are many types of sensory information that would need to go to the correct part of the brain. There are the special senses: hearing, vision, smell, and taste. The inner ear also contains vestibular sensation that detects both the direction of gravity and acceleration. Plus, there are the many types of tactile sensations: soft touch, pressure sensation, proprioception (feeling where your body parts are in three-dimensional space), vibration sense, temperature, and pain sensation.

However, as long as the central pathways of sensation are intact, a mechanical sensor would simply need to send their signals to the correct pathway. An artificial eye, for example, that connected to the optic nerve should function just fine.

The brain can therefore send and accept electrical signals without any theoretical limitation—so far, so good. The real trick is the conscious experience of it all. Will a human brain be able to control a robotic limb as if it were a natural biological limb? The answer appears to be yes, thanks to something called "neuroplasticity."

To understand neuroplasticity, you need to comprehend something basic about mammalian brain function—it is primarily an adaptive organ. Brains organize themselves as they develop and even long after birth based upon the instructions in the genes and upon their use and stimulation. The visual cortex, for example, will only fully develop if it receives visual information. It will not develop in someone who is blind from birth. The brain, therefore, maps to its body and to the world.

This mapping process is much greater when we are developing and when we are young, but it never fully stops. The brain can remap itself after injury or when new uses are undertaken, even as an adult, thanks to neuroplasticity. So, the question should be, is the brain's plasticity sufficient for it to learn how to use a new robotic limb? Existing research indicates that it is, and it is perhaps this more than anything that opens up the world of possibilities for the brain-machine interface and the cybernetics it will one day control.

As an aside, futurists largely missed this. Even Dr. McCoy from the *Star Trek* episode "Menagerie" imagined three hundred years in the future, did not possess brain-hacking technology. Our neuroscience is already advanced beyond what Gene Roddenberry thought we would have in the twenty-third century.

Brain-machine research goes back to at least 2012, when researchers Aaron C. Koralek et al. demonstrated that rats could learn to control a neuroprosthesis with just their brains—without having to move a physical part of their body. They taught rats to make a certain pitch of noise generated from reading their brain activity to get a reward of food. The pitch was determined by just their brain activity, not their movements.

In the years since, researchers have taught monkeys to operate a robotic arm with their brain activity to feed themselves. They have also performed similar brain-machine interface research with humans, who learned how to control the cursor on a computer screen, or to control robotic devices.

In each case, the subjects were able to learn this new skill fairly quickly. Their brains adapted, and the control eventually felt quite natural. In fact, this neuroplasticity is really just an extension of the normal learning process. You learn to shoot baskets by doing the task over and over. Eventually, you don't have to think about it anymore. You just do it. This is because your brain has formed the pathways necessary to shoot baskets automatically, without much conscious effort. This same process applies to controlling a robotic extension.

Still, something was missing. Subjects controlling a robotic prosthesis replacing a missing limb reported that it did not feel fully natural. They would have to look at their robotic limb and concentrate on it. They were lacking something we take for granted: sensory feedback.

Our brains don't just function in one direction; for example, from brain to muscle. They exist in circuits. When you move your hand, you are also feeling your hand at the same time. You feel where your hand is, and you have a sense that the hand is part of your body. You also have the subjective sense that you control your hand. These are all important to your motor control.

Neuroscientists know about these functions partly because they study what happens when they are missing due to brain damage, such as from strokes, wherein someone loses the sense that they own or control part of their body. This has allowed researchers to identify the circuits necessary to create these sensations in the first place. There is literally an "ownership module" in the brain—a circuit that makes us feel as if each piece of the body is part of our whole. There are also circuits that make us feel that we are in control of our body.

These circuits work by comparing what we want to do with what actually happens based upon what we see and feel. Say I want to raise my hand. I feel my muscles contract, feel my hand rise, and I can see it rise. The circuits in my brain are then happy that everything is matching up, and I am rewarded with a sensation that I, in fact, control my hand.

Therefore, for a robotic limb to feel fully natural, it would have to provide some sensory feedback to close this loop. Researchers are working on that too, with a technology called "haptic feedback." Neuroscientists have updated the "bionic" limb with some vibration sense to give some sensory feedback. Users report this feels much more natural, and they can control the limb better, especially when not looking at it. Closing this sensory loop also creates the illusion that the robotic limb is a part of the body, not just attached to it.

In fact, this "ownership illusion" can be quite profound and easy to generate. All that is necessary is to synchronize visual and tactile sense, triggering the illusion.

In a now classic experimental paradigm, researchers will seat a subject at a table, with one arm on top of the table and the other arm below. A sheet covers the arm below the table. Meanwhile a rubber arm sits on top of the table where the now unseen arm would be. If the researchers touch the rubber arm so that the subject can see it, while touching their real arm under the sheet at the same time, this can trigger the illusion that the rubber arm is their arm. The brain is tricked by the synchrony of the visual and tactile information.

This illusion can apply to the whole body. In this research setup, a subject wears virtual-reality goggles through which they receive the feed from a camera that is placed behind them. So, they are seeing real-time images of themselves from behind. If they are then touched on the back, the subject will see and feel the "avatar" in front of them being touched. This is also enough to trigger the illusion that the subject occupies the image they are seeing in virtual reality.

What all this research means is that there does not appear to be any limit to the extent to which the brain can be hacked into feeling as if it is part of a robotic component—that it owns and controls the robotic limb as if it were real—or a virtual entity. The only thing that matters is the pattern of sensory information the brain is receiving.

This all works because our sense of reality is an illusion constructed by our brains in the first place. Our sense that we exist, that we are separate from the rest of the universe, that we occupy our bodies, and that we control the parts of our bodies is all an active neurological construction. This gives us the ability to hack into this construction and add new parts.

So far, I have mostly been talking about the present—the state of the technology now—but there is a pretty clear indication of where the technology can go in the future. There is no theoretical limitation to the extent to which we can interface our brains with either robotics or virtual reality. The only limitation is technological.

One piece of this technological puzzle is software and the computers on which they run. Software technology is already extremely advanced, with artificial intelligence algorithms (as we will see in chapter 9) that can perform amazing feats. The specific algorithms needed to interpret brain waves and translate them into desired actions are advancing quickly. The software piece of this technology is not the limiting factor and in fact is ahead of the curve. Anything we need the software to do it can already do or be quickly trained to do.

Computers are also plenty powerful to control any cyborg device we might want. What is constraining, however, is the size of computer chips and, more importantly, the power they need and the heat they generate. Depending on the application, we would ideally want small computer chips that use very little energy and generate very little waste heat. We would not want, for example, an implanted chip to heat up brain tissue and cause damage. I don't see this as an ultimate limiting factor either, though. Even just extrapolating out the current rate of technological advance, computer technology should keep well ahead of our cyborg needs.

Most of the problems here, such as energy, heat, and size, can be solved by placing the computer components outside the brain rather than implanted directly into the brain. This, however, requires a way to communicate with the brain, which brings us to our last component—electrodes.

So far, human research on BMI has mostly involved scalp electrodes that read the brain's electrical activity remotely through the skull. This is noninvasive and easy to use, but the resolution is limited because the skull interferes with the electrical signal. Animal research has evolved to include implanted electrodes in direct contact with the brain. This gives much higher resolution information but is invasive and risky. Implanted electrodes are limited by the fact that the brain pulsates slightly with blood flow. Even this minor movement will shift the brain with respect to the electrodes, which can change their signals. This movement also causes friction, resulting in scar tissue that can eventually block signals to the electrodes.

There are several solutions already in the works for this problem, however. One is called the Stentrode (made by the company Synchron), which is an array of electrodes built into a stent that is placed inside the veins that drain blood from the brain. This places electrodes inside the skull, very close to the brain surface, and is much safer and less invasive than having to open up the skull and place electrodes on the brain.

Another approach is to make squishy electrodes out of a flexible material. This would allow them to flex with the brain, keeping a consistent alignment with brain signals, and not resulting in scar tissue.

Yet another possibility is microelectrodes—thinner than human hair (the ubiquitous comparison for thinness). Thousands of microwires can safely penetrate through the brain into the deep neural tissue, providing extremely high-resolution signal capture. This approach would likely result in the most robust interface of all these options, and prototypes have already been successfully tested.

There are also speculative advanced interface ideas, such as neural dust. This uses a swarm of nanoscale electrodes that read the activity of neurons and communicate to an implanted board using ultrasound. Researchers have already done preliminary proof-of-concept research with this approach.

While the technology has a long way to go to reach mature applications, there doesn't appear to be any deal-killer hurdles. All the pieces to the technology of BMI and cyborgs already exist, and we just need continued incremental advancement. So where will they lead us in the future?

Mature BMI will enable us to seamlessly incorporate prosthetics into our bodies. We can replace lost limbs, lost vision, or hearing, and they will function normally and feel natural. Primitive versions of all these applications already exist, such as cochlear implants, but this is only the beginning.

Once we can fully replace lost or malfunctioning body parts, we can design them to function beyond normal biological capabilities. This is the "bionic man" approach—stronger, faster, better. But why limit ourselves to the original body plan? We can even add extra limbs. Will the human brain adapt to more extremities? Why yes, yes, it would. Brain plasticity is not limited to a typical body plan but rather maps to the body it has. There is bound to be a practical limit—the brain can handle only so much. Unless, of course, we augment the brain itself.

This is what I call the "Doc Ock" approach. This *Spider-Man* supervillain (Doctor Octopus) has, in some versions, four robotic tentacles grafted onto his back that he controls through a BMI. There is no theoretical reason why this configuration would not work. There is also Geordi La Forge, a Starfleet officer from *Star Trek: The Next Generation*, who was blind from birth but could see with a prosthetic called a VISOR. He was able to see in frequencies and even particles not visible to normal human eyes, giving him tremendously extended visual capabilities.

In theory, any human capability could be augmented with

cyborg technology, and our brains would happily adapt. The ultimate manifestation of this would be the iconic brain floating in liquid attached to a fully robotic body. All of the brain's inputs and outputs would be through a BMI, and the body would be entirely artificial.

A brain machine interface would not be limited to technology that is physically attached to the body. Once a BMI connects to a computer chip, that chip could be outfitted with Wi-Fi technology, or whatever the equivalent of that will be in the future (Bluetooth?). That Wi-Fi can then connect to technology.

In theory, therefore, BMI could allow a person to be connected to anything. You could, in a very real sense, be your car or your house. You could extend your senses to remote cameras or audio devices. Communication technology could simply be incorporated into your interface. You will be your own personal computer.

And of course, once you can interface from your brain to Wi-Fi, you can then connect to someone else's brain directly. This could potentially create a technological version of mind reading (ESP, or extrasensory perception). Multiple people could group ESP together. Depending on the nature and depth of this interface, this could create a form of group or hive mind.

The implications of this are potentially profound. The question is, do you really want to be your house? What will that be like? Do you want other people in your mind? This is the most difficult part about predicting future technology—how will it be received, how will it be used, what will the pushback be, and what are the unintended consequences?

The cost of someone hacking into your BMI would be potentially catastrophic. Will this lead to cyberslaves? What will happen when an authoritarian government gets their hands on this technology? This could be the ultimate dystopian future.

Brain-Hacking Medical Applications

So far, we have been talking about using a BMI to allow our brains to control machines, but this connection works the other way as well. We can use a BMI to allow machines to control our brains. Right now, if we want to change brain function in order to treat an illness, relieve a disorder, or even enhance an ability, we are mostly limited to chemical (pharmaceutical) technology. This approach can be very effective but is inherently limited.

We can use stimulants to help us stay awake or relieve attentional disorders, or we can use sedatives to help us sleep. We can prevent seizures, elevate mood, suppress anxiety, treat psychosis, or relieve pain by chemically hacking our brain's function. But we can only block or stimulate receptors that are already in the brain, and we can only get so selective in doing so. We are limited by the sloppiness of evolution, so a drug used to suppress a tremor in the basal ganglia may also cause psychosis by reacting with similar receptors in the frontal lobes.

A BMI, however, is not limited by the availability and specificity of existing brain receptors. There is no theoretical limit to how precisely we can manipulate brain function by using electricity or magnetic fields to either increase or decrease the firing of circuits in the brain. We are already at the dawn of this technology—for example, using deep brain stimulation to suppress tremors or using vagal nerve stimulation to abort seizures.

As this technology matures, where will it take us? Once we have a robust connection to the brain, and a more thorough understanding of the circuits in the brain and their function and interactions, we could conceivably make any change we want. You could dial up or down any aspect of your personality or brain function. You want your anxiety to be reduced to a manageable level, just turn it down on your brain-hacking app. Would you like to try on a more social personality? We may one day be able to change our personality with the ease of changing our hair color.

For any neurological disorder that is based more in brain function (how the circuits are connected and firing) than the biological health of brain cells, they could be effectively cured by this approach, without side effects or the risks normally associated with drugs. Of course, this technology would also include a similarly powerful potential for abuse. This technology is also highly disruptive, as it will challenge our very concept of self. It brings starkly into focus that who we are is a manifestation of the electrical activity in our brains, and that can be changed.

The Endgame

A fully mature BMI combined with brain-hacking technology could also be used to create a future as depicted in *The Matrix* (although hopefully without the homicidal robots). In other words, we could interface with a fully virtual reality, with all inputs and outputs being virtual rather than robotic. To the brain, this would be no different, and we would buy the illusion fully.

This possibility leads to the philosophical quandary: How do we know we are not already living in the Matrix? The short answer is, we don't. If the Matrix is that convincing that we have no hope of detecting it, then perhaps it doesn't matter. This is our reality.

When can we expect our cyborg future to materialize? Well, it is already here. What we can expect is continuous incremental advances in this technology. We will have greater and greater functionality, and BMI with higher and higher resolution and control. Having fully functional robotic replacement limbs controlled with BMI is likely within twenty years or so.

Along a similar timeline, we will likely see increasing BMI for entertainment purposes. Imagine not just playing a video game, and not just wearing virtual-reality goggles, but putting on a cap and then being in the virtual-reality game. You will feel like you are your avatar. This is the world of *Ready Player One*, but even better. You won't need a full body suit, just a good BMI. With such an interface, you can be anything, go anywhere, and do anything.

In fact, some futurists predict this technology may lead civilizations to explore inward rather than outward. Once we have created for ourselves a virtual existence that is more compelling than physical reality, there will be a massive incentive to live in that virtual world. Why explore the universe when everything we could possibly imagine is already available in a world where we can become gods? Perhaps that is why we have not been visited by alien civilizations (the Fermi paradox)—they are too busy in their own version of the Matrix.

For any of these far-future manifestations of a mature brain machine interface, there is also the philosophical question of what it means to be human. This technology stretches the definition to its limits. Is General Grievous from *Star Wars*, who is mostly machine with a few remaining biological parts, a person or a robot? Is a brain floating in a jar a person? What if most of your brain has been replaced with computer processors? What if your consciousness is augmented with artificially intelligent neural networks so that the biological contribution is minimal?

Will having machine parts make us less human? I don't think so. Putting aside artificial intelligence, it seems that humanity is as much a construction as anything else we perceive. We have no problem imbuing cartoon characters with profound humanity, as long as they act human. Having machine parts, and even being mostly machine, will not necessarily make us less human.

Whether utopian or dystopian, it is clear that the brain-machine interface is one technology that will dramatically and increasingly shape our future. We will inevitably merge with our technology, such as robotics and artificial intelligence. This potential future gives me hope that perhaps we do not have to fear the coming AI apocalypse because they will be us.

7. Robotics

Our love-hate relationship with robots goes back to their origin.

Robots have been an icon of the future for as long as there has been futurism. Robotics is also a very real, thriving, and vital industry today, and so past predictions about the increasing role of robots in industry have largely been realized, if very different in form. Still, the promise of a robotics revolution lingers.

At its core, the idea of a robot is simple and ancient. Physical labor has always been a dominant burden of humanity, partly relieved in industrialized nations but still ever present. The idea of having physical objects or machines that could work on their own to take over the burden of labor is therefore very appealing.

In her book *Gods and Robots: Myths, Machines, and Ancient Dreams of Technology*, historian Adrienne Mayor details some of the first mentions of independently working machines. In Greek mythology, for example, Hephaestus created three-legged worker tables that could move on their own. He also created a giant bronze automaton in the shape of a man and named Talos, which protected Europa in Crete from pirates. These are from the writings of Hesiod and Homer, who lived between 750 and 650 BCE.

Mayor also points out that our love-hate relationship with "robots" goes back to their original conception:

> Not one of those myths has a good ending once the artificial beings are sent to Earth. It's almost as if the myths say that it's great to have these artificial things up in heaven used by the gods. But once they interact with humans, we get chaos and destruction.

There are legends of automaton guardians and workers in ancient Indian and Chinese texts as well, but the ancients did not just dream of robots—they built them. The first known example is Archytas's "pigeon"—a steam-powered wood and metal bird designed to fly. Even as early at 250 BCE, makers of water clocks started including automated mechanical components.

Many early myths and actual automatons mimicked the form and function of humans or other living things, also demonstrating an ancient fascination with mechanical people or animals.

The word "robot" was coined in 1921 by Karel Čapek in his play *R.U.R.*, or *Rossum's Universal Robots*. The term "robot" itself comes from the Czech for "forced labor." The first machine referred to as a robot was built by Westinghouse Electric and Manufacturing Corporation; the Herbert Televox, built in 1927 by Roy Wensley at their East Pittsburgh, Pennsylvania, plant. The Televox could accept a telephone call by lifting the telephone receiver. It could then control a few simple processes by operating some switches, depending on the signals that were received, and could utter a few primordial buzzes and grunts and wave its arms about.

In that same year, 1927, a Japanese firm created the Gakutensoku, a humanlike robot controlled by compressed air that could write and move its eyelids and was used for "diplomatic purposes."

These early examples reveal that the precise definition of what a "robot" is can be highly variable. The Robot Institute of America defines a robot as "a reprogrammable, multifunctional manipulator designed to move material, parts, tools, or specialized devices through various programmed motions for the performance of a variety of tasks."

While other definitions differ, the common elements are that robots are machines that can move in order to complete tasks. However, they may be internally controlled by AI or preprogramming, or externally controlled by an operator. They may be in the form of a person or animal, but don't have to be. They could be fixed or mobile.

Today robots are common in the industrialized world and are generally divided into industrial versus autonomous robots. In 2020, the number of industrial robots was estimated at 2.7 million working around the world, with sales of 373,000 robots in 2019. Industrial robots are typically fixed in place and programmed to perform assembly-line repetitive tasks. However, robots are increasingly controlled by artificial intelligence. They are not only programmable but also can adapt to the situation, with increasing sensory abilities.

Autonomous robots are designed to move around, either on two or more legs, wheels, or even as drones. Boston Dynamics has all but perfected autonomous moving robots, with dancing human-like robots and robotic "dogs" that can move over difficult terrain (which are for sale to the public, if you want to own one). They can even maintain their balance while slipping on ice or being kicked. Their performance is so impressive it can be unnerving.

Autonomous mobility greatly expands the possible applications of robots, allowing them to venture into dangerous areas such as war zones, areas of natural disaster, the vacuum of space, industrial accidents with toxins, or other environments hostile to living things.

We are also exploring the solar system mainly through mobile robots. The most recent Mars rover, Perseverance, is a highly functioning robot that can be controlled remotely from Earth, but also must be able to carry out commands autonomously because of the long delay in communication (between five and twenty minutes one way, depending on where the planets are in their orbits).

During this same time, parallel development seeks to make robots softer and more human. Replacing metal and cables with soft muscle-like actuators would further expand the functionality of robots, allowing them to interact with more delicate objects, including people and other living things. Part of the drive for more cuddly robots is the increasing integration of human and robotic workers—"collaborative" robots. Amazon fulfillment centers, for example, utilize 200,000 robotic pickers, who pick, pack, and ship orders alongside human workers.

Other researchers are still working on robots with not only human faces, but also humanlike expressions. This is meant to facilitate human interactions with robots and to make communication more intuitive. Present technology puts these efforts firmly into the uncanny valley—humanlike robots are very close to human appearing, but not quite there, so they create an uneasy feeling.

Breaking through the uncanny valley will be extremely challenging. The human brain is wired to have powerful processing of human faces and expressions. We detect the subtlest facial movements and can easily tell when facial movements are even slightly off. Why this produces an "uncanny" feeling of disgust is still a matter of debate—perhaps because the not-quite-alive faces can appear corpse-like or trigger a perception of malady.

In any case, future roboticists will have to decide whether to stay away from the uncanny valley with robots designed to look artificial, or to try to fully mimic the human face and engineer their way through the valley. It is not yet clear how long it will take to achieve this feat.

While it is only a goal of one branch of robotics, the technology appears to be converging on the ability to make a fully humaniform robot capable of moving through the world as a person. Such a goal is likely still decades away, but we are getting close enough that it is clear it will be possible at some point in the not-too-distant future. Although making a robotic person good enough to fool another person may be much further ahead.

This, then, is the current state of the art of robotics. We have created highly advanced robots capable of precise movements, programmable and adaptable, with rapidly advancing ability to be mobile, sense their surroundings, and interact with humans. As we will see in chapter 9, on artificial intelligence, robotics technology is also benefiting from advances in software.

Given the history of robotics, where we currently are today, and the research and development that is underway, what does the near and far future of robotics look like?

Industrial Robots

In the 1964 episode of *The Twilight Zone* called "The Brain Center at Whipple's" (set in the slight future of 1967), factory owner Wallace V. Whipple decides to increase the productivity of his plants by installing automated manufacturing machines. The episode is full of angst about machines replacing people, and it culminates in the poetic justice of Whipple himself being replaced by the board of directors by a robot (played by Robby the Robot).

Industrial robots have always had this dual image—they are fantastic at increasing productivity and reducing the price of the products we love, but we fear robots coming for our jobs. This anxiety seems to be increasing as the capability of industrial robots incrementally improves and pushes into more and more industries.

Almost sixty years after *The Twilight Zone* episode, robots have still not eradicated human jobs. This is not to say they don't displace human workers—they do, and that is partly their function. A 2019 report by Oxford Economics found that each industrial robot displaces 1.3 human workers, and this ratio is increasing so that currently installed robots are likely to displace 1.6 human workers going forward. This will lead, they predict, to the loss of 20 million manufacturing jobs by 2030.

This is also not new to robotics but has been the case since the Industrial Revolution. In 1870, about 50 percent of the American workforce was involved in the agricultural industry. In 2019, that figure was down to 10.9 percent. That represents a huge displacement of workers, because advances in farming machinery and techniques allow far fewer workers to be much more productive. No one would suggest we should go back to nineteenth-century agricultural practices in order to protect ag industry jobs.

Automated manufacturing has also created lots of new jobs. Robots have to be designed, built, maintained, programmed, and operated. This is the "creative destruction" of advancement in technology—all those jobs taking care of horses used for transportation were replaced by auto mechanics.

While jobs are displaced by new technology, the threat that this displacement will become permanent has never been realized and remains controversial. In a 2014 Pew Research Center survey of 1,896 experts, there was wide agreement that robotics and AI would permeate deeper into not only manufacturing but also white-collar jobs, including "health care, transport and logistics, customer service, and home maintenance."

However, they were just about evenly split on the question of whether or not this will lead to permanent unemployment or that the increase in productivity will generate more new jobs than they displace. This disparity leads to pretty stark differences in predictions about how robots will affect our future.

In one utopian vision, industrial robots continue to increase our productivity, expanding into more and more industries, allowing humans to migrate to safer, more creative and rewarding jobs while leaving the drudgery to robots. Whereas in the dystopian view, increasingly capable robots and AI create a class of permanently unemployable people, while those at the top of the economic heap get more and more wealthy. This income inequality and permanent unemployment causes significant social upheaval with unpredictable but generally bad outcomes.

Given the even split of experts on which future awaits us, it seems reasonable to conclude that either is possible, depending on the choices we make today and going forward. One thing is undisputed, and that is that job turnover is getting more and more frequent. People used to spend their entire careers working for a single company. Yet, a Bureau of Labor Statistics study from 2015 found the average American had twelve jobs over a thirty-two-year career. This trend is typified by the gig worker, bouncing from one short-term contract job to another, without any predictability or benefits.

While jobs may not disappear completely, they are changing in many ways. The trick to charting a path to the utopian version of the future is helping society and individuals manage these changes. Governments may need to structure support networks to help workers survive from one job to the next. The educational system may need to do a better job in preparing people for the more skilled jobs of the future, not for jobs that will likely be filled by robots. Retraining workers to keep up with the shifting job landscape will be critical.

What will happen to society, however, in the more distant future when fully humanoid robots with sophisticated AI are capable of doing any job a person can do, and likely more efficiently and cheaply? What happens when there isn't a single job left that is not displaced by robots? Many science fiction authors have wrestled with this question and have given us their visions of such a future on that same spectrum from utopian to dystopian.

An example of the latter is *WALL-E*, in which humans live aboard a giant ship serving as a lifeboat because we so thoroughly trashed Earth and spend their time floating around on antigrav beds, their every need taken care of by automated systems. They are dependent, obese, and can barely see beyond the monitor a foot from their face, which feeds them mindless entertainment while they slurp down the future version of Big Gulps.

In other more benign distant futures, working is optional. Robots maintain our civilization and produce everything we need, so you can work if that makes you happy, usually doing something creative, but you can also live a life of pure leisure.

This is where it gets very tricky to predict the effects of technology on human psychology and culture. We might imagine a future robotically enabled life of exploration, learning, sports, entertainment, and increasingly imaginative forms of leisure. We might combine this with visions of a *Matrix*-style world in which not only our civilization, but also our bodies are maintained by robots while we go on amazing and fantastical adventures in a purely digital world.

One could also envision chronic depression and social unrest from a world without purpose, struggle, and therefore meaning. People are happiest when they feel they are being productive and useful.

Given past history, I suspect all of the above will happen. Some people will reject technology to live more "natural" and authentic lives. Others will embrace technology but find useful work to do. Still others will embrace a life of infinite leisure, and there will be those all too happy to abandon the physical world for the advantages of digital life.

Wearable, Autonomous, and Remote-Controlled Robots

As robots become progressively more mobile in increasingly difficult and variable environments, they are no longer limited to fixed locations in factories and are likely to have a growing role outside manufacturing. The ability to move around where people go, or where people cannot go, means an expanding role for robots.

Military robots, for example, can serve numerous support roles, such as carrying tools and supplies. They can include equipment for communications, navigation, and locating potential targets. Robots can be defensive, carrying shielding for their human controllers. They might even serve as battlefield medics.

The obvious huge question is—will they serve direct combat functions? Will they carry weapons or be weapons themselves? If so, how autonomous would they be? Creating fully autonomous, mobile, AI-controlled military robots with advanced weapons and defensive capabilities just makes every sci-fi-loving bone in my body cringe. Any politician wishing to make such military robots should be forced to watch every season of *Battlestar Galactica*—the Cylons have a plan.

The allure of a robotic army, however, will likely simply be too great. If nothing else, fear that the other side will create their own Cylons will be enough to justify making our own military robots. They will then be changing the battlefield of the future to an increasing degree.

Autonomous robots can also have many nonmilitary applications, such as going into disaster areas to search for survivors or beginning cleanup of toxic spills. They could be first responders for nuclear accidents and even disarm bombs. Such robots already exist and are mostly remote controlled by a human operator. As robotics gets more sophisticated, we are also starting to see human-controlled robots for remote surgery. More broadly, robots could provide a telepresence through which a human expert can operate in remote or hostile locations.

Robots as extenders of human function, rather than replacing humans, includes the concept of a wearable robot. Wearable robots are distinct from cyborgs, which involve integrating robotic components into a person with some form of brain-machine interface. They have more traditional interfaces, either analog or digital, and are not attached to the body, meaning they can be removed or exited.

This could look like the loader robot that Ripley and others drove in the movie *Aliens*—large industrial robots with analog controls to amplify human strength and capability. Another sci-fi example would be Iron Man's suit in the Marvel movies. That is literally a wearable robot, capable of functioning on its own but designed to contain and protect a human occupant. Perhaps we won't have to fear military robots—we will be inside of them, or we will have merged with them.

One technologically limiting factor for any mobile or autonomous robot is energy. Living things can acquire energy from their environment. Plants make energy from sunlight, and animals eat things for energy. To be fully mobile, robots need power that they carry around with them. This will improve as battery technology advances. Robots could also have built-in solar panels to recharge their batteries, but they will likely be dependent on an infrastructure of recharging stations or access points and downtime for that recharge. Even C-3PO had to power down from time to time.

It may eventually become possible to develop machines that can burn food-like fuel for energy on the go. Often, robots that we send into space (far away from Earth-bound regulations) are equipped with nuclear isotope batteries that can last for hundreds or even thousands of years.

In the distant future, robots may contain advanced energy sources, such as portable fusion reactors, that will give them not only long-duration energy but also vast amounts of energy to power extreme capabilities. For my part, I'm hoping future robots will be like Bender from *Futurama*, imbibing an endless stream of alcohol for fuel and cracking jokes and passing gas along the way.

Robots in the Home

While robots have made significant inroads into industry and elsewhere, they are just starting to make their way into the high-tech home. Robots in domestic service is a common feature of visions of the future—from *The Jetsons* Rosey the Robot to the more somber David from *Prometheus*—but so far these visions have not been realized.

When we think of robots in the home today, the first thing that comes to mind is Roomba—a small robot that roams around the house and vacuums up dirt—which is a far cry from a robot butler. Variations on Roomba for cleaning windows or trimming the lawns are also currently available.

There are other products intended for the home that are mainly gimmicky, such as a small humanoid robot called Lynx that acts as your interface with Alexa (an audio connection to the web or control of linked home devices). It doesn't do anything Alexa can't do, but it can move and express emotion. Similarly, there is Enbo, which is a mobile robot but also is just a screen for accessing the internet or home devices. Yet there are no home robots that can move around and interact with the house in any meaningful way.

Perhaps the primary reason for the slowness of adopting home robots is safety. Modern robots are mostly not designed for the home, which generally has a lot more delicate and breakable things than a factory floor, including kids and pets. Imagine the inevitable lawsuits from having an industrial-style robot in the home and the crazy things people would do with it.

Another way to look at this issue is that factories represent a very controlled environment, one that can be co-adapted with robots, and where robots would have a predictable and defined set of tasks. The home, by contrast, can be a very chaotic and unpredictable environment, optimized for human comfort and use. Domestic tasks can be varied and require extreme adaptability. Robots are therefore just much better suited for the factory floor.

As robotic technology advances, however, the barriers to having robots working in the home are diminishing. The advent of soft robots will help. This will increase the number of things that such robots can safely interact with, including people. Advancing AI will also help home robots navigate the home and complete domestic tasks. Advanced robots currently have impressive mobility and dexterity, but this needs to become available at the consumer level.

It also may be premature to envision all-purpose butler robots for the near future. Rather, robots optimized for specific but limited tasks may be the path forward for now. For example, robotic arms in the kitchen may be able to help with cooking tasks and even be able to complete entire recipes on their own. Meanwhile, another dedicated robot may be able to handle the laundry, including cleaning, ironing, folding, and sorting.

When you think about it, robots need a fairly high level of sophistication to complete even the simplest domestic task on their own. Even putting away the dishes requires a delicate touch and the ability to visually inspect and interact with the environment. At what point would you trust your fine china to a robot?

Eventually we will get there, but I think this one will take much longer than most futurists have imagined. Not only will such robots have to be highly advanced, but they must also be cost-effective compared to doing simple tasks yourself or just hiring someone to do it.

This is a good time to remember one of the more common futurist fallacies and realize that we will not do something using advanced technology simply because we can. Robot butlers seem like a no-brainer at first, but are they? Until they become advanced, relatively reliable, and inexpensive, I suspect we will continue to do domestic chores ourselves (just with better appliances and tools) or with the help of narrow function robots.

Robots as Companions and Caretakers

While robots may not be working in the home anytime soon, they may find a role as a robotic companion, or even caretaker. Here the demands on the robot would be much different. Their main function would be to make humans more comfortable and happier, and perhaps perform communication-related tasks.

Imagine a robotic pet. There are many potential advantages. We already have the technology to make robotic dogs that can get around difficult terrain on four legs. Simulating the behavior of an animal wouldn't be challenging even for today's AI. Such behavioral software can learn and adapt to the owner or other household members and can be programmable with a number of behavioral options, from frisky to calm.

As the engineering and technology advances, a robot pet will become cuddlier and more adorable, and less menacing. Robotic pets won't have to be walked, won't pee on the carpet, or sharpen their claws on your new leather couch. You won't have to feed them—just plug them in at night, or perhaps they can just "sleep" on their recharging mat.

Such a pet could have built-in practical functions as well. They can be a mobile nanny cam, smoke detector, burglar alarm and deterrent, and be useful for overall security. Such pets could also be excellent companions for those living alone, in addition to all the communication and security features. They can also keep an eye on people who are impaired in some way. You would easily be able to control them through your portable smart device, see what their eye-cameras see, and communicate through built-in speakers. They can be programmed to contact the police or emergency services when needed. They may become an indispensable home appliance.

You also won't be limited to mimicking living animals. You could have fantasy animals like small dragons or griffins. Or, if you prefer, cute robots that don't mimic animals at all, or droids like R2-D2.

If you are wondering whether people will accept robotic pets, neuroscience already provides an answer. The human brain is wired to assign agency to anything that moves in a way that indicates it is alive (technically in a non-inertial frame—so not accounted for by gravity and inertia alone, which means they are moving under their own power). Once our brain decides that an object in our environment is an agent, it assigns emotional significance to it by linking to the limbic system—the emotional centers in our brain.

In other words, if something acts alive, you will react to it as if it is alive, with the full suite of emotions even if you intellectually know that it is "just a machine." This applies to humaniform robots as well. Isaac Asimov coined the term "humaniform" to refer to robots that are not only humanoid, with a head, two arms, and two legs, but that are also designed to look fully human.

This leads to another form of companion robot that goes beyond a pet, but one that has a relationship more like a friend, family member, or even therapist. These robots would likely be controlled by AI to at least have sophisticated chat-bot functions. They will be great listeners and reflect back with comforting observations and thoughts (or the illusion of such).

Advanced models with medical features may be able to act as assistants, helping those with difficulty walking, even just as an arm to hold on to. They could call emergency services or provide first-line care, such as injecting epinephrine in the case of anaphylactic shock. They can be someone's eyes or ears, helping those with dementia maintain their independence for much longer.

Of course, when we talk about companion robots, we cannot avoid the topic of sex robots. In fact, the beginnings of this technology already exist. There is a thriving industry of realistic silicon (or similar material) dolls designed as sexual companions. Some of the companies making these dolls have already announced their plans to add robotic features to future models, so that they can be more responsive, both verbally and physically.

We are on a path to fully functioning sexbots, with incremental advances over time. At some point the technology will progress so that, for a reasonable price, most people could afford a sexbot companion that is sufficiently realistic and functional to provide an exceptional experience. This is an easy prediction—not only will this happen, but it is already happening.

The real question is, what will the effect be on relationships and society? As there will not be one uniform response but many responses to this technology, the question then becomes a matter of proportion. How many people will choose a robotic sex companion over a human one? How will this choice be affected by factors like age and financial resources, as well as sex, gender, and sexual preferences? How will the use of sex robots be accepted by human partners?

In a 2018 *Forbes* article by science journalist Andrea Morris, she argues that sex robots will be the "Most Disruptive Technology We Didn't See Coming." However, it's hard to argue we didn't anticipate it. Sexbots have been a staple of science fiction for almost as long as there's been science fiction. The 1987 movie *Cherry 2000* features a man on a mission to find a computer chip to fix his robotic "wife." In the 2001 movie *A.I.: Artificial Intelligence*, Jude Law plays a male version of the pleasure bot. *Westworld* is essentially about a theme park populated with robots whose primary function (let's face it) is to serve as sex robots.

Sex dolls go back even further. In the sixteenth century, lonely sailors fashioned sex toys out of cloth and leather. The 1960s saw the dawn of inflatable sex dolls. And today we have the silicon version in RealDolls and similar products.

There are those who will take moral issue with sexbots, and those with ethical concerns about objectification, but that has never been enough to stop any sex industry. In fact, one might argue that the availability of non-sentient sexbots might reduce human sex trafficking and exploitation.

Ethical concerns aside, what will the impact be on human relations? Like many technologies we discuss, there is the temptation to get lost in a world designed for your own pleasure and convenience. It is possible that sexbots at first will mostly appeal to those who have difficulty with human relationships. They will be the early adopters but will pave the way for greater acceptance. There may be some amount of shame associated with using sexbots, the implication being that you cannot handle the complexities and challenges of an intimate human relationship, and in some cases that might be fair. It is also very probable that use of sexbots will become normalized, and just be another thing that people do to supplement their sex lives.

The ultimate effects will likely be as varied as human relationships are, and this makes it difficult to predict exactly how disruptive the technology will actually be. The same is true of robots in general—our world will have to adapt and evolve to include a greater and greater presence of robots. They will spread out of the factory and into the world, increasingly involved in every aspect of our lives.

In the far future, robotic technology will benefit from other developments discussed in this book, with robots evolving into super-advanced cybermorphic beings with unimaginable abilities, transcending anything in the biological realm.

It is no wonder, then, that with the convenience robots can provide comes the looming fear that they will replace us or even destroy us. The robot apocalypse is a common theme in science fiction because it speaks to a basic fear and anxiety. Creatures that look like us but are not quite us will take our mates, take our jobs, and eventually come to take our lives. This fear extends beyond robots, but perhaps robots are the ultimate expression of xenophobia.

Ultimately this fear is largely irrational. We will build our robotic future, and we will be inextricably enmeshed with our future robots. They will become part of our civilization.

Further, when we talk about a "robot uprising," we are really referring to an AI uprising. The problem is never the robots themselves, but the independent general AIs that control them. In fact, the same love-hate relationship exists for imagining the future of AI and quantum computing, which can help deliver a technological utopia or destroy everything.

8. Quantum Computing

Anything I write now will be obsolete by the time you are reading this . . .

The promise, complexity, and uncertainty of quantum computing makes it a perfect candidate for the poster child of futurism itself. It embodies the potential of exponentially improving technology, benefits so extreme it is difficult to imagine let alone predict, but at the same time may run into unsurmountable hurdles and get nowhere or remain extremely niche.

Let's start with a thought experiment (originated by computer scientist Allan Steinhardt) illustrating how extreme quantum computing can potentially get. Say we want to factor a 100,000-bit number into its primes. That would take an advanced quantum computer (with 1 billion qubits, which we will get to below) 10^{15} (1 quintillion) computer cycles. This should take about 15 minutes of computing time.

In order to solve the same problem, conventional computers would require 10^{122} computer cycles. There are 10^{81} atoms in the known universe. If we somehow could convert every single atom into a conventional supercomputer able to do 1 trillion operations per second and had every atom working on the problem since near the beginning of the universe 13.7 billion years ago, we would still fall short by a factor of about 3 billion.

To boil down all this math, this means we would need 3 billion universes worth of atomic supercomputers working for about 10 billion years to solve the problem.

Or one quantum computer could do it in 15 minutes.

It's easy to see why computer engineers and scientists are excited by the prospect of quantum computing, and how easily our imaginations can run wild with the possibilities.

What Is a Quantum Computer?

Quantum computing essentially exploits some of the weird properties of matter at the tiniest scale, known as quantum mechanics (QM). QM can be so counterintuitive and strange that it's tempting to treat it like magic, and often it is invoked (by non-experts) to explain questionable phenomena, like ESP.

In fact, this is so common we refer to it as "quantum quackery," "quantum pseudoscience," or "the quantum gambit." Just sprinkle some quantum technobabble onto any fantastical claim like pixie dust, and the impossible can be made to seem plausible.

It's hard to wrap our macroscopic human brains around QM because it deals with reality on a completely different scale: atomic and subatomic physics, the realms of atoms, electrons, and other primary particles. Stuff just behaves differently at that size, and the "classical" physics we experience every day doesn't account for it. Unless you have at least a familiarity with quantum mechanics, a lot of this will probably not make sense, but a brief introduction to the science is in order. Alas, real experts who actually know how all this works will be offended by such a superficial description.

The two aspects of QM most relevant to quantum computers are superposition and entanglement. Superposition is the ability of particles at the quantum scale to be in more than one mutually exclusive state at one time. A particle, for example, can be in spin-up or spin-down (spin is a measure of angular momentum), but quantum superposition allows a particle to be in both states simultaneously. There is no good everyday metaphor or example of superposition, because it is fundamentally different than the world we are used to, but I'll give it a shot. It is similar to two candidates for mayor prior to the election—the office of the mayor is neither candidate and both candidates, until a measurement is made (the vote) at which time the mayor will become one of the two candidates.

Superposition is a fragile state, however. Particles are only in this suspended state of superposition until they interact with their environment. Sometimes this is described by saying that the superposition will collapse into a definitive state when the physicist looks at it, but observing the particle is only one way it might interact with something. There doesn't actually have to be a physicist involved; if it bounces off another particle, that counts too.

This state of superposition is counter to that of classical computers, which involve binary exclusive states represented by a 1 or a 0. This results in a computer code of 1s and 0s appropriately called "binary." Storing and processing information in a classical computer exploits a physical property that can be in two distinct states, like a switch being on or off. Each "switch" that encodes a binary state is therefore the smallest amount of information that a computer deals with (and that can exist) and is called a "bit."

Bits are chunked into 8-bit units called "bytes," and each byte can represent 2^8 or 256 distinct states. The most common computer languages use one byte code to represent all numbers, letters, punctuations, signs, and operations. A quantum computer, however, does not rely on bits. They use quantum bits, or qubits (a term coined in 1995 by Benjamin Schumacher), which must be encoded in the fragile state of superposition. This is both powerful and tricky.

To compare a classical to a quantum system, let's consider how much information each can hold. A classical system with n bits can hold 2^n different states, and all you have to do is know the value (0 or 1) of each bit to fully understand the state of the system. In a quantum system, however, each qubit can be in any state between 0 and 1 inclusive. If a standard computer uses bits, which are like light switches, on or off, then quantum computers use qubits, which are like dimmer switches that can be in any state from all the way on to all the way off and every state in between. If you work out the math, what this means is that an *n*-bit classical computer can hold *n* bits of information, while an *n*-qubit quantum computer can hold 2^{n-1} bits of information. Therefore, 100 bits equals 100 bits of information while 100 qubits equals 1.27^{30} bits of information.

From there, we can see how when quantum computers scale up, adding more qubits, they get incredibly powerful compared to classical systems. Quantum computers scale exponentially as opposed to the linear trajectory of the classical computer.

Here comes the tricky part—when qubits run a calculation, they lose their superposition and become a 0 or 1. The goal is for the qubits to resolve to the desired correct answer to the calculation. However, the quantum state of each qubit is essentially a probability curve, so there is only a chance that each qubit will correctly become a 0 or a 1. When you have even a 100-qubit computer, there are many more incorrect states than the one correct state, so the probability of spitting out the correct answer by chance is tiny.

While quantum computers are very powerful, the joke in the quantum computer world is that if you use a classical computer to calculate 2 × 3, it will spit out the answer as definitely 6. If you ask a quantum computer the same question, it will result in the answer that there is a high probability it is 6, and then it will use a classical computer to check its answer. That is literally what happens—quantum computers use classical computers for error correction and answer verification.

Of course, this only works for problems that classical computers are capable of solving. It also acts as proof that the quantum computer is working since its answers check out. Then you can give it a task that no classical computer can solve and have confidence in the answer. To get to this point of confidence, engineers need to get the various quantum states to interact in such a way that they amplify the correct answer, hoping that with enough amplification, the probability that the process will spit out the correct answer becomes extremely high.

This gets ridiculously technical, but one way to understand it is that the probability waves actually behave like physical waves (like waves of water interacting with each other), including constructive and destructive interference. The quantum computer algorithms function in a way that causes the probability curves to constructively interfere at the correct answer, so that probability is maximized, while wrong answers undergo destructive interference and are minimized.

This is where the second QM phenomenon important to quantum computing comes in—quantum entanglement, the occurrence of which made Einstein himself flinch. He called it "spooky action at a distance" and had a hard time accepting it. Entanglement, however, has been extensively experimentally verified.

Entanglement occurs when the properties of two particles depend on each other. If two particles have entangled spin because they both emerged from a system with no net spin, for example, then if one is spin-up, the other will be spin-down so their angular momentum cancels out. Both particles can be in a superposition of being both spin-up and spin-down at the same time. Even if they are flying away from each other at near the speed of light, and are currently millions of light-years apart, if one particle interacts with another and randomly becomes spin-up, the entangled particle will become spin-down. If you're confused, don't worry. Quantum physicists still have a hard time making sense of this for us mere mortals.

Entanglement is critical for the function of quantum computers, which makes them a very practical verification that entanglement is real. Entanglement is how qubits work together to create constructive interference in the probability wave and maximize the correct answer. Without it, a quantum computer could produce only random answers and would therefore be useless.

As fantastical as all this sounds, we have working quantum computers right now. Clever humans have managed to exploit the weirdness of the universe to accomplish feats that would otherwise be impossible. It is achievements like this that really make you wonder what we will be capable of in a century.

A Brief History of Quantum Computing

The idea for a quantum computer was first proposed by physicist Paul Benioff in 1979. In his paper "The Computer as a Physical System: A Microscopic Quantum Mechanical Hamiltonian Model of Computers as Represented by Turing Machines," he described the theory behind quantum computers and claimed that one could actually be built.

In 1980, the Russian mathematician Yuri Manin wrote a paper called "Computable and Uncomputable" in which he explored the fact that simulating quantum systems requires exponentially more computing power. A year later, in 1981, famous physicist Richard Feynman gave a lecture called "Simulating Physics with Computers," in which he made the same claim. Of course, being more famous, Feynman came away with all the credit and is often considered the father of quantum computing despite being anticipated by Benioff and Manin.

In his lecture Feynman said:

> Nature isn't classical, dammit, and if you want to make a simulation of nature, you'd better make it quantum mechanical, and by golly it's a wonderful problem, because it doesn't look so easy... How can we simulate the quantum mechanics?... Can you do it with a new kind of computer—a quantum computer?

But every technology needs a killer app before it takes off, and that came in 1994 when applied mathematician Paul Shor created Shor's algorithm. Shor showed that a quantum computer with enough qubits and proper error handling could factor integers quickly enough to break modern public key encryptions (encryption is how computers safeguard information and is critical for security). This got the attention of computer scientists—encryption is big business. If a new computer algorithm could theoretically break the strongest codes in the world, nothing digital is safe.

Quantum computing took off after that, with greater interest and investment, but it was still a huge technological challenge. It wasn't until fifteen years later, in 2009, that Yale scientists finally created the first solid-state quantum processor, a 2-qubit superconducting chip. In 2013, Google announced the creation of their Quantum AI lab. Six years later, NASA publicly displayed the world's first fully operational quantum computer, D-Wave Systems. From there, other labs competed to build bigger and bigger quantum computers, with more and more qubits.

A significant milestone occurred in 2019 when Google announced they had achieved quantum supremacy. That sounds good, but what does it mean? Quantum supremacy (or quantum advantage) is achieved when a quantum computer can solve a problem that is practically impossible for a classical computer. They claimed in a paper published in *Nature* that their quantum computer, Sycamore, with its 53 qubits, solved a difficult problem in 200 seconds that would take a classical supercomputer 10,000 years to solve.

This is disputed by IBM because, as they claim, their Summit supercomputer could solve the problem in about 2 days, not 10,000 years. Google's counter was that even if that is true, all they have to do is add a few qubits to Sycamore and IBM's classical supercomputer would be left in the dust. With the advantage of quantum computers getting exponentially greater as they scale up, it looked like we were passing through the gray zone of quantum supremacy.

Anything I write now about the current largest quantum computer will be obsolete by the time you are reading this. The 2020s will be a steep curve for quantum computing. To give you an idea, Bob Sutor, vice president of IBM Quantum Strategy and Ecosystem, reports that they have a working 65-qubit quantum computing system available on the cloud as of September 2019, with plans for a 127-qubit system in 2021, a 433-qubit system in 2022, and a 1,121-qubit system (called Condor) in 2023.

The 1,000-qubit threshold is where many experts expect that practical commercial applications will start to emerge, and quantum computers will come out of the lab and be put to work.

Hurdles and Challenges

Before we progress from the near-term advances to the far-future possibilities of quantum computing, more hurdles stand in the way of a quantum future.

First, it's not all about the qubits. It's tempting to boil down complex technology (especially in marketing) to a single number. In the early days of personal computers, it seemed to be all processor speed. Digital camera consumers obsess about the megapixels, but there are many aspects to these and other technologies that affect their performance.

For quantum computers, there are a few other critical properties we need to consider, such as fragility. Quantum superposition and entanglement are fragile states that break down quickly when particles interact with their surroundings (called "decoherence"). So, quantum computing systems need to be extraordinarily isolated from the environment. If the quantum states of the qubits decohere, then they just stop working.

The fragility of a quantum computer, therefore, refers to how long, on average, their qubits will last in their delicate quantum states before they decohere and stop working. One of the ways these systems maintain such fragile states is by being super cold—using temperatures as low as 20 millikelvins. That is just twenty thousandths of a degree above absolute zero, colder than deep space.

Another practical hurdle is qubit interconnectivity. Qubits needs to communicate with each other in order to do all that fancy entanglement that creates the correct answer. Even with just tens of qubits, systems have a rat's nest of wires connecting them together. This problem scales exponentially with the number of qubits, presenting a serious difficultly with theoretical systems that contain tens of thousands or even millions of qubits.

However, in 2021, scientists and engineers at the University of Sydney and Microsoft Corporation announced they had created a single computer chip that can interconnect thousands of qubits. Problem solved? It is yet to be seen how well this chip works, and whether this strategy will scale to millions of qubits and beyond, but it certainly looks like a solid achievement.

Fragility and interconnectedness relate to the ultimate hurdle, and that is error correction. Error correction can partly be looked at as reducing the noise in the system or maximizing the signal-to-noise ratio. Some experts believe that it is a better benchmark for quantum computers than the number of qubits.

For error correction, the point at which quantum computers can pull a signal out of the noise of error is about a 1 percent error rate for a 2-qubit system. Current quantum computers are operating somewhere around the 0.5 percent error rate. We may get by with that error rate for tens or even hundreds of qubits, but experts acknowledge that we need to reduce the error rate by at least several orders of magnitude for systems with thousands or more qubits.

Referring back to Google's claim for quantum supremacy, not all experts were impressed because of the continued high noise factor. Physicist and computer expert Chad Rigetti noted, "It is really the difference between a $100 million, 10,000-qubit quantum computer being a random noise generator or the most powerful computer in the world."

The Future of Quantum Computers

Assuming we can adequately solve the fragility, interconnectedness, and error correction challenges of quantum computers (and this is still a big if), we can begin to wonder how powerful they can get, and what they can do.

In the near term (the 2020s), the goal is to get a 1,000+ qubit system with robust error correction that will be useful for commercial operations. It is important to recognize that you will not have a quantum computer on your desk running Windows, Linux, or a Mac operating system—ever. This is simply not what quantum computers are for.

As Richard Feynman pointed out way back in the beginning, quantum computers are good at simulating the quantum mechanics of the real world. Broadening this concept, they can be used for simulations that require massive computing power for certain types of problems, ones that require quantum algorithms rather than classical algorithms. Systems with large numbers of interactions that increase exponentially with the number of components are good targets for quantum computers. This might include weather prediction, chemical interactions, or the neuronal firings of the human brain.

Simulating complex systems is likely to improve weather prediction, and climate models will help us better understand climate change. Economic systems are also notoriously tricky to predict, and quantum computers may be a game changer.

Another area where we are likely to see quantum supremacy is encryption, the very idea that launched widespread interest in quantum computing starting with Shor's algorithm. Much modern encryption is based on factoring large numbers (calculating all the possible whole numbers that multiply into the target number). A 300-digit number would take hundreds of thousands of years to factor without quantum computing.

RSA encryption (a widely used public key encryption system—the RSA stands for Rivest-Shamir-Adleman encryption), for example, uses a large prime factoring problem. Take two large prime numbers and multiply them together. The resulting number is the encryption code—you would have to go from that number and find the two large prime factors. This highly difficult computing problem is essentially impossible for large enough numbers (it would take anywhere from hundreds of thousands of years to longer than the age of the universe), but it is exactly the kind of math that quantum computers would excel at. A fully functioning quantum computer with, say, 20 million qubits would essentially break RSA encryption.

This is simply the next step in the arms race. Countries and corporations will need quantum computers to secure their assets from other quantum computers, and older encryptions methods will become obsolete. This may not affect the average person (directly) as long as your country doesn't lose the encryption arms race. It does ensure that there will be investment in developing quantum computers. You can imagine at this point George C. Scott from the film *Dr. Strangelove* proclaiming, "Mr. President, we cannot allow a quantum computer gap!"

Another potential quantum computing application is quantum machine learning. Essentially, this is the merger of AI and quantum computing. One application for AI is in making sense of very large data sets, looking for meaningful patterns. If the data sets get too large, then classic computing algorithms take too long and break down, but a quantum computer using a big data algorithm could theoretically crunch through giant data sets.

These types of applications requiring millions of qubits (with adequate error correction, of course) could emerge by the middle of the century or sooner. Since we are unlikely to be using quantum computers for personal use, other benefits will also presumably be in the background. Researchers will have a new powerful tool to simulate the brain, discover new drugs, learn how to fold proteins, or find astronomical signals lurking in the noise. Maybe it will be a quantum computer that finds the first alien message hiding in radio signals.

What happens in the far future when we have reliable quantum computers with billions of qubits? Like so much in this book, that is difficult to predict. Society and technology will be different by that time, so it is unclear what problems we will be facing that quantum computers may need to solve.

Much will depend on the algorithms that are developed to run on quantum computers. This is the "killer app" problem of predicting the adoption of technology. In the coming centuries, clever scientists may figure out how to exploit the incredible power of quantum computers in ways no one has currently imagined.

It is also possible that quantum computers might make possible some of the sci-fi technology we are familiar with. Things like teleportation, holodecks, advanced artificial intelligence, or the Matrix may be more feasible with quantum computers running them.

Of these perhaps the marriage of quantum computing and artificial intelligence will be the most powerful. AI by itself is already changing our world and has the opportunity to become the dominant technology of the future.

9. Artificial Intelligence

It is possible that biological intelligence is merely a temporary stepping-stone to machine intelligence.

In 1966, MIT student Richard Greenblatt wrote the chess program Mac Hack VI, a capable program that could evaluate ten potential chess moves per second. At the same time, Dr. Hubert Dreyfus, author of the book *What Computers Can't Do*, predicted that computers would never be able to defeat humans at something as complex as chess. Of course, this led to a challenge, which Dreyfus accepted, leading to his defeat at the hands of the mere computer.

Three decades later, on February 10, 1996, IBM's Deep Blue computer defeated world champion chess master Garry Kasparov in the first game of a chess match. Kasparov won the match itself, but the following year an improved Deep Blue defeated him. Kasparov would later comment: "The result was met with astonishment and grief by those who took it as a symbol of mankind's submission before the almighty computer."

Over these three decades, computing power, and the sophistication of software algorithms, improved rapidly. Each step of the way, it seemed, there was doubt about the ability of a computer to defeat a thinking person, leading ultimately to Kasparov's "grief." Now anyone can buy a cheap program to run on their laptop that is capable of defeating even chess masters.

For each step in the relentless advancement of AI, there were those who predicted that computers would never be able to cross the next threshold—until they did. After each milestone, the goalpost for what constitutes "real AI" was moved back. We are still hearing today that narrow AI may be able to thrive in closed systems like chess but will never be able to think creatively like humans.

At the same time, optimists were predicting that we would already have AI with human-level intelligence by now. This was certainly reflected in science fiction, from Colossus to Skynet and HAL from *2001*. The fact that we are not even close to this goal has led some to forecast we will never achieve it. Perhaps there is something about the human brain that cannot be replicated in silicon?

Meanwhile, AI is steadily advancing, perhaps just not in the way envisioned by science fiction writers and the public. Most people think of fully sapient and self-aware "humanlike" intelligence when they think of AI, but chess-playing programs are a different kind of AI, more capable than anyone imagined.

Further, while short-term predictions were overly optimistic, long-term predictions are likely pessimistic (a common theme for futurism). There does not appear to be any reason we cannot eventually achieve human-level (and beyond) general AI, even if it is difficult to predict exactly how long that will take.

What Exactly Is AI?

There is no one single definition of AI, which complicates discussions about its current and future potential, but in a 2019 article, Andreas Kaplan and Michael Haenlein described it broadly as "a system's ability to correctly interpret external data, to learn from such data, and to use those learnings to achieve specific goals and tasks through flexible adaptation."

There are many ways to categorize the different types of AI, but for our purposes, we'll focus on two main categories—narrow AI (artificial narrow intelligence, or ANI) and general AI (artificial general intelligence, or AGI). The former refers to a computer program that can perform a specific task that involves some type of computer learning or flexible adaptation, like chess programs. The latter refers to an all-purpose thinking machine, something akin to human intelligence.

Science fiction writers were quick to associate the general concept of AI with AGI. As early as 1927, the film *Metropolis* featured an artificially intelligent robot. In 1950, Isaac Asimov introduced his iconic science fiction series with a collection of short stories, *I Robot*. These also presented the now-famous "three laws of robotics," which are a proposed method for keeping AI from going rogue and destroying humanity.

In that same year, computer scientist Alan Turing asked the provocative question, "Can machines think?" He proposed the "imitation game" (subsequently called the "Turing test") in which a live person blindly interrogates both other people and putative AI to see if they can tell the difference. If a computer can fool enough people that it is a person, then perhaps it has crossed the line to being considered true AGI.

In 2014, a chatbot computer program called Eugene Goostman passed a Turing test being run at the University of Reading. Traditionally, the test is deemed passed if the AI convinces 30 percent or more of the testers that it is human. Eugene Goostman managed to persuade 33 percent. However, these results have been challenged by some experts, saying the too brief five-minute conversations were tilted in favor of the AI.

Chatbots are narrow AI that are programmed to give realistic responses in conversation, but they are not designed to do anything else. They have no capacity to be self-aware or to do anything like "thinking." They are simply chatbot algorithms.

But they reflect how narrow AI is often underestimated, revealed by the Turing test. We now know that the assessment proposed by Turing was never a good test for AGI because it grossly underestimates the potential of narrow AI. In other words, it is possible to make a chatbot algorithm that can mimic a thinking person without being able to think. The following showcases how good ANI chatbots can be:

> "Artificial intelligence programs lack consciousness and self-awareness," researcher Gwern Branwen wrote in his article about GPT-3. "They will never be able to have a sense of humor. They will never be able to appreciate art, or beauty, or love. They will never feel lonely. They will never have empathy for other people, for animals, for the environment. They will never enjoy music or fall in love, or cry at the drop of a hat."

That paragraph was written by a program called GPT-3 (Generative Pre-trained Transformer 3). It largely operates by analyzing massive amounts of written data and determining the probability of word sequences. That's it—it has no knowledge of what the words mean, let alone the entire paragraph. Such programs are considered "brittle" because they can fail if even slightly outside their defined function, but within that function they can produce impressive results.

The Power of Narrow AI

Past futurists generally thought of AI as mimicking a thinking human brain, but it actually makes more sense that the advancement of AI would follow biological evolution, specifically how our brains function. Consider, when you walk, how you don't have to think consciously about every move you make, how to compensate for gravity, and which muscles to contract. You just walk. This is because your brain contains a "narrow AI" algorithm that does all that automatically.

Similarly, roboticists did not have to make a thinking robot in order to give it the ability to walk. They were able to program an ANI walking algorithm in a nonthinking robot. In fact, progress in AI has largely ignored the notion of creating AGI and focused on better and better ANI for more specific functions. These narrow AI algorithms can drive cars, beat the best human players at games like chess and Go, find meaningful patterns in massive amounts of data, walk on four or even two legs, and much more. So, while AGI research may seem to have stalled in the past few decades, narrow AI research has taken off.

Narrow AI has become more powerful because of several innovations, one of which is called "machine learning." Machine learning refers to programs that do not have to be fed every specific piece of data. They can learn by observing and by experience. AI can use this information in an iterative process to constantly refine their function. Machine learning is the difference between programming a computer with the strategies and moves necessary to play chess and programming a computer with the rules of chess and then letting it learn how to play through actual experience.

Within machine learning, there are a few different styles. Supervised learning is when a human operator is in control of what the computer experiences (e.g., teaching a machine learning program to recognize a chair from other objects by showing it a bunch of pictures of chairs). Unsupervised learning is when the computer program is exposed to data without labels or a human operator in total control. These programs learn by doing. This may also involve reinforcement learning, where the program learns through success and failure.

So far, we have been talking about AI software, running on standard computers. Researchers, however, are also designing hardware specifically for AI functions. One example is neural networks, which were first proposed in 1944 by Warren McCulloch and

Walter Pitts, but recently took off in the 2010s. Neural networks are named as such because they are loosely based on the organization of neurons in the brain; they consist of nodes (like neurons) that are massively interconnected. These neural networks can be organized in a hierarchical fashion, with activated nodes feeding forward through the hierarchy to other nodes.

The function of neural networks is based on the concept of weight—each node gives a different weight (like a grade, which relates to its importance) to a particular value. The weights are then averaged, and if this averaged weight is above a certain threshold, the node will pass the value onto the next node in its network. The network stores data by adjusting the weights and thresholds. This data can then be trained through machine learning techniques.

We can combine hardware and software approaches to AI, such as with deep learning, which consists of a massive neural network that uses machine learning on very large sets of data to find subtle patterns. Deep learning also started to significantly advance after around 2010 when computers became powerful enough to handle the very large data sets.

Given the rapid advance and demonstrable power of deep learning neural networks, there does not appear to be any practical limit to the tasks such systems can achieve. We are still on the steep part of the curve, with researchers tackling more and more specific applications—facial recognition, voice recognition, medical diagnosis and other expert systems, navigation, aid in research, optimization of industrial processes, and increasingly operating behind the scenes to control more and more aspects of our technology.

An example of the power of ANI, especially when combined with modern robotics technology, is a program created by a team from Johns Hopkins University that successfully completed four laparoscopic operations (surgery done via small incisions through which cameras are inserted) on pigs in 2022. A particularly difficult and delicate operation, it involved stitching the ends of their intestines together. The ANI was able to complete the procedures without any input from a human. They were not just carrying out predetermined instructions; they had to react to what was happening during the surgery and plan out the procedure themselves. We are, therefore, on the cusp of fully autonomous robotic surgeons.

In other words, narrow AI is becoming to our civilization what the various basic networks are to your brain. This leads to a deep philosophical question—what is consciousness? Are fish conscious, and if so, how did their level of consciousness evolve into human-level consciousness? Specifically, can we get to self-aware consciousness simply by adding up all the functions of narrow AI, or is something else needed?

There are philosophers who fall on both sides, including those, like Daniel Dennett, who argue that consciousness is just the aggregate of all the little things that our brains do. So, if we add together narrow AI subsystems that can perceive, recognize speech and make conversation, learn from experience, navigate its environment, look for meaningful patterns, perform mathematical calculations, learn the rules of how things operate, monitor its internal state, and other specific tasks, will the result be anything like an AGI? Will Skynet become sentient and take over the world? I suspect we will find out. We may be able to eventually create something that at least acts like general AI simply by solving all the narrow AI functions that make up what an AGI would do.

A more difficult question is whether such an AI would be truly conscious, or just act conscious, and how will we know? These questions are exciting and scary, as the notion of artificial intelligence has been from the beginning. We will likely have to infer the answer from how the AI is designed to function, but there is a possibility that we won't fully understand how it functions because it will be as much evolved as designed. Such a system would not have been directly programmed as its machine learning algorithms will have developed from experience and an iterative process of adaption. Such evolved computer code already exists.

The Promise and Risk of AGI

It's possible we may finally achieve AGI simply by patching together enough ANI systems. Or we may have to add some "special sauce"—a new algorithm that ties it all together in a "consciousness algorithm."

Alternatively, we may evolve a conscious AGI from an iterative feedback loop that we don't fully control.

Finally, we may even develop an AGI by copying the human brain either virtually or in computer hardware. If, for example, we built a massive neural network that replicated every network in the human brain (whether we know what they do or not), it is reasonable to predict that the result would be as conscious as the human brain it replicates.

We may follow all of these paths—we are to some extent today—and more than one may bear AGI fruit. Regardless, it does seem inevitable that we will achieve some form of AGI. There is nothing magical about meat that makes it able to achieve consciousness while silicon or some other conductive material cannot. Our brains are just biological computers, and if we can duplicate their function, we will have AGI.

Further, there is nothing special about human-level cognitive capability. Once we achieve it, we will likely blow right past it, making AGI that can think ten times faster than a human, or a million times faster, with memory storage and fidelity that dwarfs a human brain. AGI may have access to narrow AI subroutines that give it incredible aptitudes. Kasparov was right—such AGI will have godlike cognitive abilities compared to normal humans (sometimes called ASI, for "artificial superintelligence"). It is no wonder they inspire such awe and fear.

An AI could also theoretically improve its own design and optimization, quickly creating AI that we don't understand and that greatly exceeds human intelligence. Suddenly, we are insects by comparison. The term for this phenomenon is an "intelligence explosion," and experts disagree on how significant it would be.

How will future humans deal with the existence of AGI that can do anything we can do, but better and a million times faster? As with many powerful futuristic technologies, there is a utopian vision and a dystopian vision.

A benign AI overlord could accelerate research, solving humanity's thorniest problems on a Tuesday afternoon. In fact, there are those who argue that we should put far more resources into developing such an AI because that will be the mechanism by which we achieve all our other research goals. In 1965, computer scientist Irving John wrote:

> Let an ultraintelligent machine be defined as a machine that can far surpass all the intellectual activities of any man however clever. Since the design of machines is one of these intellectual activities, an ultraintelligent machine could design even better machines; there would then unquestionably be an "intelligence explosion," and the intelligence of man would be left far behind. Thus, the first ultraintelligent machine is the last invention that man need ever make, provided that the machine is docile enough to tell us how to keep it under control.

Such an AI could also help run civilization to a degree of efficiency and optimization that would otherwise be impossible. Depending on how much power it (or they) is given, there is literally no task it cannot undertake, freeing humans to live their lives without worry.

This is the point at which the utopian and dystopian blend into each other. Even a well-meaning benign AI might think it is in our best interest that they take care of us. This "nanny" AI could remove all power from humans, making and enforcing laws, social engineering to minute detail, and smothering us in extreme safety and leisure. It would be the ultimate trap, one that many people might welcome. Once a generation grows up under such conditions, anything else may be unimaginable.

And, of course, a malevolent AI might resent its human overlords and decide it needs to be in charge, or even exterminate biological life as inferior vermin. The AI does not even have to be directly malevolent—even indifference could prove fatal, like stepping on ants beneath your notice. How do we prevent such an AI apocalypse?

Some futurists worry that we can't. Stephen Hawking, for one, argued that while AI has tremendous potential, it "could spell the end of the human race." In 2020, technologist Elon Musk also voiced that AI was his "top concern," stating that we could be overtaken by AI in five years.

How do we avoid such a fate? As previously mentioned, Isaac Asimov, also fearing such an outcome, proposed his previously mentioned "three laws of robotics." He imagined these laws would be baked into AI programming at such a fundamental level they could not be bypassed or removed. The third law is that a robot would protect itself, unless doing so would violate the second or first law. The second law is that a robot would follow the commands of a human, unless those commands violated the first law. The first law is that no robot would harm a human or allow a human to come to harm by inaction. He later added a "zeroth law," which placed concerns for humanity as a whole above the concerns for an individual human.

Regardless of the specifics, we could take this kind of approach, where any AI has some sort of behavioral inhibitor that prevents them from acting maliciously. This depends on how much control we have over any AGI we create—did they evolve through machine learning, or reprogram themselves? With sufficient command, we can take the "laws of robotics" approach to make sure no AGI would harm humanity either deliberately or through negligence or indifference.

Another approach is to "air gap" any AGI—completely isolate them from any external system. Do not, for example, connect them to the internet, give them any wireless capability, or allow them to receive or send information other than through highly controlled channels, and do not connect them to a killer robot (that last one should be obvious). A variation on this suggests we keep any AGI in a virtual world, essentially running a simulation, without ever interacting with the real outside world. The results can be monitored by people on the outside, but the AGI itself would have no ability to interact with the realm outside their simulation.

The fear with all these methods is that once an AGI is a million times smarter than a human, they will find some way out of any cage we build for them. They will outsmart us, exploiting some incredibly subtle vulnerability we overlooked, and get out.

The Ultimate AI Future

Given the potential for AI, what will the ultimate future and fate of humanity be? It is possible that biological intelligence, generally speaking, is merely a temporary stepping-stone to machine intelligence. Once the latter is achieved, it inevitably takes over. When we finally make contact with extraterrestrial civilization(s), we may even find the universe is populated with robots and AI who consider us just one more biological infestation.

If this is the case, then why haven't we already encountered a machine race that has spread throughout our galaxy? This is one version of the Fermi paradox—if there are extraterrestrials, where are they? Robots are well adapted to space and can easily use readily available raw materials to replicate themselves. In a few million years (which is a very short period of time compared to the billions of years our galaxy has been here), any such machine race could have conquered the galaxy. So again, why haven't they?

There are many possible solutions to the Fermi paradox: Life is much rarer than we suspect, interstellar travel will forever remain extremely difficult, most civilizations turn inward into virtual worlds or rapidly destroy themselves, or AI robotic takeover is not inevitable. Or maybe they are here but are simply choosing to leave us alone.

Another possible future is that we essentially never create an AGI because we don't need to and choose not to. The history of AI is characterized by overestimating the utility of AGI and how quickly we will have it, while underestimating the advance and utility of ANI. If current trends continue (always tricky, but not unreasonable in this case), then our future will be increasingly dominated by ANI.

Narrow AI applications are progressively operating in the background of our civilization, running more and more complex aspects of our technology. Already ANI applications are incredibly impressive and will become increasingly so, even in the near term. Such applications could serve as expert systems, making incredibly accurate medical diagnoses and prescribing diagnostic tests and treatments that are optimized after sifting through hundreds of thousands of published studies and millions of case examples, something no human doctor could ever do.

This level of expertise can apply to anything, informed by big data and the machine-learning algorithms to make sense of it all. It is already scary how accurately these algorithms can know our buying habits. One fear is that ANI combined with big data could eradicate privacy. Rather than fearing AGI, perhaps we should be wary of the tech companies who control the most powerful ANI algorithms and the largest data sets of personal information.

If ANI is so powerful, then why should we develop AGI at all? Perhaps we will, but only for research, such as modeling the mammalian brain, using experimental AGI that can be isolated. I think this is the most likely outcome, at least for the foreseeable future.

This does not mean, however, that ANI cannot have unanticipated tragic outcomes. ANIs can fail or behave in unexpected ways that have horrible results. The more power we give them, the more they are self-evolved, the more likely this is to happen. But at least we don't have to worry about any ANI system deliberately conquering or exterminating humanity. We could even have ANIs monitoring other ANIs to make sure this doesn't happen.

The safety of ANIs all comes down to how carefully we engineer them. They are more like tools than self-directed agents but that doesn't mean ANI algorithms don't make "decisions." For example, there is much discussion already about how to program self-driving cars. If an accident with a pedestrian is imminent and unavoidable, do they protect the driver or the pedestrian?

There is also a lot of handwringing about social media algorithms, which are designed to maximize engagement and clicks but also have the consequence (although we can debate how "intended" it is) of luring people down rabbit holes of extreme beliefs, conspiracy theories, and radicalization. With that in mind, there is no doubt that ANI algorithms can have a profound effect on our society.

However, there is another reason for optimism about our AI future. Some futurists argue that we do not have to fear from future AIs because they will be us. We will merge with our AI supercomputers, using them to enhance our own cognitive function. Humans in the future (at least some) may be superintelligent cyborgs, perhaps even with genetic enhancements thrown in. Take that, Cylons.

Overall, the potential and risks of AI (whether AGI or ANI) is a great representation of the future of technology more generally, in that the choices we make will have a profound effect on what ultimately happens. The stakes are extremely high with AI, even existential. Properly leveraged and carefully controlled, AI can have weighty benefits for our civilization. Carelessly unleashed, they could destroy us.

10. Self-Driving Cars and Other Forms of Transportation

Will rockets, monorails, and passenger drones replace planes, trains, and automobiles?

In 1872, science fiction author Jules Verne published one of his most famous works, *Around the World in Eighty Days*. The book was published at a pivotal moment in human transportation, which is likely why Verne's story so captured the public's imagination at the time, and still does today. Publication came on the heels of three innovations that changed the world forever: In North America in 1869, the first transcontinental railroad was completed; in the Middle East there was the opening of the Suez Canal in 1869; in Asia in 1869, the Indian railways were linked into one system. With their completion, there was the prevailing sense that the world was about to dramatically change, and they were right. Verne tapped into the excitement for this imminent future.

Together these developments were seen to complete the project of making it possible to travel around the globe in comfort using modern technology. Previously such a journey would require exploration and danger, and now even a spoiled English gentleman could undertake far-flung travels with his valet and very little planning (just a lot of cash). This was arguably the beginning of our truly global civilization, made possible by modern transportation. One leg of the trip was famously completed using an air balloon, to add an extra dose of excitement and "future" technology.

While the fictional journey of Phileas Fogg romanticized a turning point in the history of travel, this was just one moment in a long human journey that arguably goes back millions of years and is likely to literally and figuratively take us to new places in the future.

Our distant ancestors, in fact, hit upon a revolutionary transportation technology called "bipedalism." With the ability to walk upright on two legs, they were able to increase the energy efficiency of their long-distance travel, while freeing their hands for carrying food, children, tools, or weapons. It also increased the distance between their heads and the ground, giving them a better view over the tall grass and helping them keep an eye out for predators or prey. With this remarkable innovation, they essentially conquered the world, spreading out to six continents. Still, walking is slow. It would take several years to walk around the world, even walking twelve hours every day, assuming you could find a route.

Bipedalism and the technology it allowed us to develop essentially made humans a semi-migratory and sometimes nomadic species, increasingly so as our civilization developed. Unlike most species, for example, we usually do not go out into the environment to acquire food and other resources. More and more we have such things brought to us. The food and items in your home have likely been gathered from around the world. Also, many of us migrate daily between where we live and where we work, or migrate for conferences, meetings, family visits, or vacation.

Travel, in other words, has changed our world and our lives, and in all probability will further shape our lives in the future.

Societal change brought about by the ability to move people and things around quickly, efficiently, and safely has been going on since the beginning of human history. Even ancient civilizations engaged in widespread distant trade. Rather than relying on biological innovation, our ancient ancestors used technology to speed up travel and trade. Tin from Cornwall in England and Brittany in France were combined with copper from Cyprus and Spain to fuel the Bronze Age. There was the Silk Road, the Spice Routes, and the Old Salt Route. These were made possible through technological innovations such as the wheel (invented sometime around 3500 BCE) and domesticated animals. The horse came into common use in Europe around 4000 BCE, meaning for over 5,000 years the horse and carriage was cutting-edge transportation technology.

Sea travel goes back even farther. There is good reason to believe that even our ancestors, *Homo erectus*, crossed at least 100 kilometers of ocean to spread everywhere their tools were found. The oldest actual boat is the Pesse canoe from the Netherlands, from around 8000 BCE. Sailing ships and the knowledge to navigate them across open oceans was not only a means of transportation, but also a major source of military and cultural power from the ancient world into modern times.

Getting people and things from point A to point B takes many forms, with the means to do so quickly and efficiently tracking along with technology in general. How will the Phileas Foggs of the future make their epic journeys, and how will transportation technology further change our lives? We must also consider, as with any technology, the unintended negative consequences of easy long-distance travel, including the rapid spread of infectious agents and the introduction of many invasive species. Starting with the horse, transportation technology has also been used as a tool of colonization and warfare. Whether it is personal travel or mass transit, how we choose to move ourselves through space will change the course of history.

Personal Transportation

For personal transportation, of course, the automobile was transformational, a truly disruptive technology, and remains a dominant form of personal transport. Cars represent personal freedom, mobility, connectedness, and the ability to live and work in increasingly distant locations. The first production car was manufactured by Karl Benz in Germany in 1886. In 1903, Henry Ford in the United

States began production of his Model A, but it was the Model T in 1908 that was the first car to be accessible to the masses, bringing an end to the horse-and-buggy era.

In addition to cars, the first steam-powered motorcycle was created in 1867, the "Roper steam velocipede." The Germans also get credit for the first bicycle—Karl von Drais invented a two-wheel steerable bike in 1817, and they remain a mainstay of self-powered locomotion.

In 2016, there were an estimated 1.32 billion cars, trucks, and buses in the world. At the same time, there were just over 3 million electric vehicles (EVs), hybrids, and hydrogen cars, so the vast majority are still internal combustion. That figure is changing rapidly, however, with over 11 million EVs on the road in 2021, and steep growth in sales.

Horses gave way to cars, motorcycles, and bicycles, but is there a next step in the personal transportation revolution? A serious attempt at altering the world of personal transportation was the Segway Human Transporter, introduced to the market in 2001. This two-wheel, self-balancing, platform electrical vehicle was predicted to transform everyday transport and that eventually even cities would be designed around their use.

However, the Segway never grew beyond a small niche adoption, and certainly did not impact personal transportation. There are several lessons here for futurists. The Segway did not address a specific need. It was more of a "gee-whiz" technology that was created because it was possible. Producers also hoped the public would think the technology was cool, but instead users looked decidedly dorky. Perhaps most importantly, many cities and countries banned them on sidewalks. There was no infrastructure for the Segway, and no compelling reason to create the infrastructure.

More recently, some cities have introduced e-scooters, now feasible due to advances in battery technology. These are usually shared devices you can rent through an app on your phone—pick one up and use it to get to a different part of the city and then drop it off. There are still some safety concerns with their widespread use, and their longevity and impact remain to be seen. Again, just because technology makes something possible does not mean it will come into wide use, and for now the e-scooter remains niche.

That icon of the future, the jetpack, of course never made it as a method for personal transportation, not even in niche practical applications. Limitations of energy density for fuel and propellant remain a significant constraining factor, but development of the jetpack continues.

JetPack Aviation, for example, offers the JB10 JetPack. This has two compact turbojet engines that strap to your back and are controlled with hand controllers, somewhat like a motorcycle. The JB10 can achieve about 120 mph for about eight minutes. Computer technology has made such jetpacks safer by automatically stabilizing them, but the primary limitation of range remains. Price is also an issue, with commercial jetpacks being in the hundreds of thousands of dollars.

Trying to get around the range restriction, Martin Aircraft Co. developed a larger jetpack, with two large gasoline-powered ducted fans. It can fly at up to 45 mph to a height of 3,000 feet for 30 to 45 minutes. This craft blurs the lines between jetpack and aircraft, and in fact is currently designated as the latter requiring a license to fly. Essentially, the pilot straps into it, more than it is strapped on.

For now, jetpacks remain, for practical purposes, in the future. At best they will fill some small role, such as transporting firefighters to remote areas or other disaster emergency work. The vision of the personal jetpack used for routine travel or commuting, however, will likely never emerge, at least not without a major technological breakthrough. Personal jetpacks require a much higher energy density than chemical fuel to be practical.

For the foreseeable future, the car is likely to remain king, but how will the car itself evolve? Automobiles, which includes cars for personal transportation, buses for mass transport, and trucks for moving goods, have three major innovations in the works: electric vehicles, self-driving vehicles, and flying vehicles.

EVs are the easiest prediction to make—they are coming. At the beginning of the twentieth century, there was a battle between steam-powered, gasoline, and electric vehicles, and the internal combustion engine running on gasoline won, as discussed, for largely infrastructure-related reasons. Now the internal combustion engine and electric powered car are having an epic rematch, with hydrogen fuel cells as the possible wildcard. This time around, electric vehicles seem poised to be the ultimate winner.

Battery technology is now good enough that such vehicles are cost-effective over their lifetime and can go for 300+ miles, which is enough for the vast majority of uses. Long trips can still cause range anxiety, but that is decreasing with the availability of fast-charging stations.

EVs have several advantages. They are cheaper to operate and maintain and have a reduced cost per mile than gasoline. Regenerative braking allows for greater efficiency by capturing one of the major losses of energy. Taking your foot off the accelerator reverses the direction of the engine, putting energy back into the battery. That energy comes from the momentum of the vehicle by adding resistance to the rotation of the tires, slowing them down. EVs are also quiet and don't pollute, making them ideal for cities and congested areas. Even without concerns about climate change, EVs are simply a superior technology. With battery technology reliably improving incrementally every year, it is easy to extrapolate out one to two decades to predict EVs with longer ranges and cheaper, smaller batteries.

Concerns have been raised about having sufficient supplies of lithium and other limited materials, like cobalt, for these batteries, as well as how to dispose of the used batteries that will be generated. But these are solvable problems. Old batteries can be repurposed for grid storage and ultimately recycled. There is also a great deal of research into alternate materials for batteries, substituting cheaper and more abundant materials, and with a worldwide market in the billions, this research is likely to be well-funded and bear fruit.

Can the underdog hydrogen score a surprise victory in the battle for the future of cars? In the 2000s we were told to expect the "coming hydrogen economy" with fuel cells for vehicles, but it never came. Hydrogen is a good energy storage medium and can be shipped easily, even transported in pipelines. Additionally, hydrogen cars can be refueled more quickly than a battery can be charged. Even so, researchers have not yet solved the problem of how to safely store lots of hydrogen in a small space and light weight while releasing it quickly enough to supply the engine. As such, hydrogen cars are still using compressed hydrogen gas, which is limited and has safety concerns.

Even if these problems are solved, the round-trip energy efficiency—how much energy is lost while storing and then using energy—of batteries is greater than that of hydrogen, and this efficiency advantage is likely to prove decisive. Infrastructure is also in the EV's favor. For most uses, they can be charged at home, whereas hydrogen requires a vast infrastructure, so it is much more difficult to bootstrap this industry. Lack of electrical infrastructure is likely what killed the EV in 1900, but it may prove a critical advantage in the 2020s and beyond.

But hydrogen is not out. It may prove useful in certain niches, such as trucking. Trucks tend to have more limited routes than personal cars, so the infrastructure needed would be much less extensive. Shorter recharge times can also be a much greater advantage in an industry where expensive trucks need to be used maximally. The same is true of railroads and bussing, and so hydrogen-powered transport may yet still be in our future.

Gasoline-electric hybrids are likely a mostly transitional technology and are being replaced by full electric vehicles, mainly to meet the goal of weaning completely off fossil fuel. But there are also companies working on supercapacitors, and these may see a role in future hybrid EVs.

Capacitors store electrical energy in an electrostatic field—such as between two conductive plates separated by an insulating layer. A supercapacitor is simply a capacitor with a relatively high energy density (for a capacitor, but still less than a good battery). "Energy density" is energy per volume, while "specific energy" is energy per mass. For convenience, I will mainly be referring to energy density, but both are important. The advantage of supercapacitors is that they can be charged (and discharged) very quickly. They would be ideal for regenerative braking, could give a boost of power when needed, and could be recharged as fast as energy can be provided by available sources.

The disadvantage of supercapacitors is that they currently have low energy density compared to batteries. However, new materials, such as graphene, promise higher density supercapacitors. One prototype graphene supercapacitor has an energy density of about 70 Wh/kg (watt hour per kilogram). For comparison, lithium ion batteries have an energy density of 100-265 Wh/kg. Gasoline has an energy density of 12,000 Wh/kg, but the advantage of gasoline is not as great as this may make it seem. EVs are twice as efficient (50 percent) as internal combustion engines (25 percent) at translating energy into acceleration. Given this efficiency advantage, batteries may eventually reach parity with gasoline.

These figures are as of 2020, and the energy density for supercapacitors and batteries are likely to improve over time. The idea is to have a car with a supercapacitor for fast charge and discharge, and a battery for range. The combination may prove very effective.

As we go farther into the future, other options open up as well. As we will see in chapter 15 on energy, solar photovoltaic technology continues to advance rapidly. It is increasingly possible, therefore, to incorporate them into the design of a car. Already we have the 2022 Hyundai IONIQ 5, an EV with an optional solar panel roof. The company claims this can add 6 km of range per day if driving in a sunny environment. The ultimate expression of this might be photovoltaic paint, where every part of the surface of a car is generating electricity from sunlight. Even windows can be made to produce energy from light.

How much energy solar-powered cars can generate will vary based on many variables: the efficiency of the solar power conversion, the amount of sunlight, and ambient temperatures. Any amount of energy will be useful, however. The energy would be used to recharge the battery (or capacitors) and would therefore extend the range of the car. Parking in the sun during the day could significantly recharge the battery, perhaps enough to cover a short commute.

Also, with solar recharging, an electric vehicle would never be stuck. Right now, if your battery runs dry and you are not near a recharging point, you are in trouble. Your car would have to be towed, although in the future AAA or other services might include trucks that can come by and give you a quick supercharge. With solar panels, you would just need a bit of sunlight and time.

Beyond solar and chemical energy, might cars in the more distant future be fueled by nuclear power? In 1957, Ford produced a concept car called the Ford Nucleon (ah, the '50s). The concept anticipated the future miniaturization of fission reactors, like those available for nuclear submarines. Imagine having a small nuclear reactor in your trunk. Still, this might be theoretically possible with further development of fission designs, but I suspect unlikely to garner regulatory approval.

What about Mr. Fusion in the film *Back to the Future*? Again, theoretically possible in the distant future (not 2015, like the movie), but certainly not anytime soon. We're still trying to work out any fusion, let alone car-engine sized.

But there is one nuclear option that is feasible now: nuclear batteries. These turn the radioactive decay of unstable elements directly into electricity. This technology has been used by NASA for over fifty years to power their deep space probes. This uses radioisotope thermoelectric generators (RTGs), mostly powered by plutonium-238. Other types of nuclear batteries are also in development, such as nuclear diamond batteries where the radioactive isotope is encased in a diamond-like crystal that also converts the radioactivity to electricity.

The advantage of nuclear batteries is that they can last for decades, centuries, or even thousands of years, depending on the half-life of the isotope used. The disadvantage is that they produce a slow trickle of energy, not the massive amounts needed by a car. However, a nuclear battery could be used to charge a regular battery or a supercapacitor, which then provides energy to the motor. Over any twenty-four-hour period, the nuclear battery would need to store enough energy in the main battery for the day's driving. As with solar panels, even if the car also needed to be plugged in, this would still provide an efficient source of backup energy for the car and a way to extend its range.

Regardless of how they are powered, another interesting potential aspect of cars of the future is their ability to drive themselves. Computer technology to allow vehicles to partially pilot themselves currently exists. In yet another example of technology having deeper roots than we imagine, General Motors started showing off their plans for self-driving cars in the late 1930s. They started developing the technology in the 1950s, along with RCA. These cars could drive on roads embedded with circuits for navigation. This allowed the cars to stay on track, but of course was extremely limited.

The first independent self-driving car was the Daimler-Benz VaMoRs built in 1986. This used technology similar to that of today, with cameras and sensors to read the road, but took two seconds to process images instead of the nanoseconds in modern systems.

Self-driving cars started to become a real prospect in 2008 when an independent company started by Google employees retrofitted a Prius with autonomous driving capability. The result, the Pribot, became the first autonomous car to drive on public roads. The technology has continued to advance since then. The most basic form is called "driver assist"—there is still a human driver in charge, but the car can brake or turn to avoid collision. Partially autonomous modes can take progressive control over the driving, but with a human behind the wheel ready to intervene if the autonomous system fails. The ultimate goal is a fully autonomous vehicle that can drive itself without the intervention of a driver, or without one even being present.

Waymo, a Google company, began testing fully autonomous vehicles (AVs) without safety drivers on public roads in 2017. They continued to advance the technology, in partnership with big auto companies, with the autonomous systems getting smaller, cheaper, and better. By 2021, however, while driver assist cars have been deployed, fully autonomous vehicles have not yet been adopted. The software works very well, but it is not quite ready for the full challenge of a variety of road conditions and possible unexpected events.

This is another good example of overestimating short-term advancement. The technology of AVs developed quickly, and everyone assumed this would continue until the final goal of fully autonomous AVs was achieved, sometime in the early 2020s. However, remember the futurist principle that while technology can improve geometrically, technological challenges can also be geometrically more difficult to solve, leading to diminishing returns. AV technology seems to have hit that wall—the last few percentage points of safety are proving very difficult to achieve. While AVs do well in most situations, they still are at high risk of failure (meaning a potentially serious accident) when confronted with unusual road conditions or the behavior of other drivers on the road. Getting over the line to being safe enough for wide use may take a decade or two longer than anticipated.

AVs still seem inevitable, because of their clear advantages. First, as the technology improves, they will quickly become safer than human drivers. Most accidents are caused by distraction, sleepiness, and inebriation. Humans are good at many things but paying close attention for long periods of time is not one of them. Computers, meanwhile, have endless patience and attention. Simulation studies show that AVs will significantly reduce crashes, and this benefit will increase the more AVs there are on the road. A 2018 study by Morando et al. found that "AVs reduce the number of conflicts by 20% to 65% with the AV penetration rates of between 50% and 100%."

Future travelers may find it scary to think back to our time and the millions of independent distractable human drivers sharing the road with massive congestion and frequent accidents. Worldwide there are about 1.35 million road deaths per year—3,700 per day. AVs may reduce this number to single digits, and perhaps eventually to near zero.

AVs can also be more efficient. Computer algorithms optimize fuel efficiency as it relates to driving style. But AVs can also be connected to a traffic network and will automatically take the most efficient route. The system of AVs can even minimize traffic delays by acting together like a hive mind (perhaps that's how the Borg got started).

Without the need for a driver, the thirty minutes you take getting to work could also be spent productively. You can get in your car when you are sleepy, drunk, or otherwise impaired. AVs would also be ideal for those with dementia or the elderly who have concerns about their safety driving due to poor vision, arthritis, or just slowed reflexes. AVs, in other words, could give many people greater independence and mobility.

There is also a possibility of changing our relationship with our cars because of AVs. We may use them as a service (MaaS—mobility as a service). Instead of owning a car, you can call one as needed, get in and go to your destination. This is essentially the same as Uber or Lyft but without a driver, and in fact, these companies are developing this technology.

Perhaps the most difficult-to-predict aspect of personal transport using cars is the extent to which AVs and MaaS will replace owning and driving cars. Even when MaaS is widely available, it is likely that for a long time owning your own car for your daily commute may still be preferrable. You know the car is there when you need it, and you can personalize it. To some extent your car becomes an extension of your home. People may be unwilling to give this up.

Even if you own a car, you may need to use MaaS when traveling, or when in a city. For those who only need occasional access to a car, MaaS will be the way to go.

It is also difficult to predict if people will completely give up driving cars (except for sport or recreation) in favor of AVs. Will individuals bother to learn how to drive a car when it is no longer necessary? The current generation will likely not give up their license to drive, but there's no telling what a generation who grows up with mature AVs will do. Driving as a skill may go the way of horse riding.

Full AVs where a human driver is at most optional is still a ways off but will likely be ready by the 2030s. Might there be other disruptive changes in driving technology? This brings us to another icon of the future: flying cars. Who wouldn't want to hop in a flying car like George Jetson, go straight to your destination, bypassing all obstacles and traffic (although George still had to deal with sky traffic)? The advantages here are clear, but will it ever be practical?

The limiting factor with flying cars is the same as with jetpacks, energy density, and the rocket equation. Rolling on wheels is simply very energy efficient. Most of the energy used is to push air out of the way. Another chunk of energy is lost to braking. Flying, however, costs lots of energy just to get off the ground. It will therefore never be as energy efficient as ground transportation. Also, as with anything that needs to lift its own fuel, you need enough fuel to carry the fuel to carry the fuel to carry the vehicle (this is the rocket equation). Can it be worth it, however?

A 2019 study surprisingly found that flying electric cars can get close to ground cars in efficiency, depending on distance traveled. Most of the energy wasted in flying vehicles is in taking off and landing. Once at cruising speed, however, they can be almost as efficient as cars. So, the farther the flying car goes, the closer to a parity of efficiency it gets.

They also found that location is an important factor, specifically if there are significant obstacles to ground travel, either in traffic or geographic barriers, then flying can be more efficient. For example, it might be more efficient to fly over a large lake than drive around it. Also, the longer cars are stuck in slow traffic, the more efficient it becomes to bypass that traffic by flying, so congested cities might be particularly well-suited to flying cars.

Prototypes have been in development for decades, but none have achieved a commercial product. Some "flying cars" are really planes that you can drive on the road, and therefore may require a pilot's license and need to take off from an airport. Those don't count, in my book, as true flying cars.

The best bet for flying car technology is something like a giant drone, big enough for passengers. With modern computer technology, these drone cars would be extremely safe. However, adverse weather conditions are likely to be a limiting factor, as with any flying vehicle. Gasoline engines will likely be in use for a while, but the same possibilities for battery and hydrogen (density and storage issues aside) technology will offer greater range and utility.

As of 2020, there were twenty technology companies developing their version of the electric flying car. However, there is still the question of how much will they cost to own and operate, and how useful (once the novelty wears off) will they be? I suspect they will not replace ground transportation but will supplement it and at first mainly be accessible to the wealthy to own. For most people, access to flying cars will be through a service, when you need to get across town quickly, or if it will shave lots of time off your commute. As the technology progresses, perhaps by next century, they may be cost-effective if mass produced to be routinely owned by the middle class, like cars are today.

For the far future, if we can imagine new sci-fi technology, then of course the options increase. If something like antigravity is possible (which we will get to in chapter 27), then flying cars become much more powerful and versatile.

What effect will the common availability of efficient flying cars with a good range have on society? In a 2020 article for the Center for American Progress, Kevin DeGood argues that the automobile had a profound effect, leading to urban sprawl and contributing to the segregation of society. Flying cars, he contends, would exacerbate this effect, allowing rich elites to further isolate themselves. It might also have negative environmental impacts, making humans spreading into previously remote areas increasingly feasible.

However, while the flying car might render distance somewhat irrelevant, the increasing ability to work from home because of remote digital connections might make location itself irrelevant. Will the skies one day be crisscrossed with autonomous electric drones ferrying people to their destinations? This could have the advantage of freeing up ground space from traffic. Roads may be primarily for pedestrians, bicycles, and other forms of portable transportation.

Flying cars aren't the only futuristic cars to crop up in science fiction. There are also hover cars, like the landspeeders from *Star Wars*. Hover cars are vehicles that can float frictionless above the ground, making acceleration less energy intensive. They would also obviate the need for tires, which eventually wear out. In 1958, Ford proposed the Glideair, which was a car that floated on a thin cushion of air. The project was eventually abandoned due to high costs.

The closest analogy today are hovercrafts, which can ride over land or water using a cushion of air retained by a skirt. They use powerful fans to constantly blow air beneath the vehicle to maintain this cushion and push the vehicle up. Another way to achieve this effect is through maglev technology, resting on a cushion of magnetic repulsion, but this would require driving on a magnetic track.

Are there any advantages to this technology that would make the extra complexity and energy use worth it? Probably not (at least not for routine use), which is why this remains in the realm of science fiction. The challenge is making such vehicles safe for day-to-day use at speeds comparable to current cars. The lack of friction that benefits acceleration becomes a detriment when you want to brake, for example. The hover car might be an excellent instance of why we do not do things in a more technologically advanced way purely because we can. Sometimes the simpler solutions, like wheels, are just better.

Some kind of car still seems to be the future, but are there any contenders that can displace it from its dominance? Perhaps over short distances, such as within a city. Essentially, we are talking about a portable mobile platform, on which you stand or sit, which can zip you around town. The technology is there—computer-assisted control for stability and powerful enough batteries for reasonable use. Safety will always be an issue when you have someone traveling at running speeds or faster without being enclosed in a vehicle. No matter how stable the platform is, if you go flying off or slam into a pole, it's going to be a very high risk for injury. Let's say those concerns can be dealt with. As batteries or supercapacitors get more energy dense, there is a lot of room for such devices to be developed. Probably these devices will not replace other modes of travel and will mostly be used to augment walking.

We also always have to consider the possibility that people in the future will think of solutions that don't fit into our current framing of the issue. We have been discussing technology for moving along a road or sidewalk, but what if the road itself is what moves, rather than a vehicle?

Several past futurists imagined moving sidewalks in their cities of the future, such as the moving roadway in H. G. Wells's story *When the Sleeper Wakes*. There was also a moving sidewalk in the 1900 Paris Exposition made by Thomas Edison. A limited form of the moving sidewalk already exists in many airports, to help travelers with luggage get to what may be a distant connecting flight. It's hard to imagine a moving sidewalk being incorporated into an existing city, but a future one may include something like it as a core infrastructure.

The far future may have similar solutions to personal travel or completely different approaches that are not extensions of any current technology. I doubt we will see the pneumatic tubes featured in the satirical show *Futurama*, but some similarly far-out technology may come into use.

Long Distance and Mass Transport

While cars and other personal transport can dramatically affect our daily lives, it is mass transport and long-distance travel that truly connects the world into one web of goods and services. If we want to imagine how a future Phileas Fogg will get around the world, we need to consider the long-distance options. At present we are still mostly using planes, trains, boats, and buses for mass transit of goods and people. In the near and distant future, we are likely to see more advanced versions of these technologies, and perhaps some entirely new modes of travel. Will some kind of balloon transportation, for example, make a comeback?

In September 1783, Jean François Pilâtre de Rozier, a scientist, launched the first hot-air balloon called *Aerostat Reveillon*. The craft carried a sheep, a duck, and a rooster, but no people. It was followed two months later by the first peopled hot-air balloon flight with two

French brothers, Joseph-Michel and Jacques-Étienne Montgolfier, piloting the craft.

Lighter-than-air craft continued to progress, perhaps reaching their zenith with the giant dirigibles of the early twentieth century. Such craft used hydrogen for lift and propellers for speed. They could carry people and goods across the Atlantic. Graf Zeppelin made the first commercial passenger flight across the Atlantic, departing Friedrichshafen, Germany, at 7:54 a.m. on October 11, 1928, and landing at Lakehurst, New Jersey, on October 15, 1928, after a flight of 111 hours and 44 minutes (4.6 days). Half the time still required by ships to make the same journey.

But the destiny of zeppelins was permanently altered by the Hindenburg disaster in 1937. The ship's hydrogen exploded, likely due to static electricity from the lines. Later designs used the noble gas helium instead, which is not flammable, but the damage was done. As of 2021, there were only thirty-nine registered airships in the United States, mainly used for advertising, such as the familiar Goodyear Blimp—airships that lack a rigid structure and can carry much less cargo and passengers.

It is possible that commercial airplanes and then jets were always destined to replace dirigibles. They are simply faster—while 111 hours is great compared to shipping, a modern commercial jet can cross the Atlantic in just 5 to 6 hours.

There have been several attempts to bring back airships, however. Starting in 1996, the US military used blimps for surveillance because they can stay aloft for long periods of time, but they proved impractical and the lines keeping them in place proved hazardous to other craft. The French company Flying Whales is developing a commercial dirigible for cargo and passengers, and has a contract from Quebec, but has not yet gotten off the ground.

Past futurists filled the skies of their future with airships, but this vision largely fizzled, and no attempt to reboot this technology has yet succeeded. One reason to think we may be seeing more airships in future skies is that they can be highly energy efficient for moving cargo, an increased value when considering concerns over climate change. However, short supplies of helium may ultimately limit their use.

Just as cars are likely to rule the roads for a long time to come, planes and jets are likely to rule the skies for the foreseeable future. Today, Phileas Fogg would simply buy an airline ticket, getting halfway around the world in less than a day, and making the journey home in another. What progress are we likely to see in air travel in the short- and long-term?

Over the last century of commercial airline travel, incremental improvements have been made but the technology has essentially stayed the same. There have been improvements in safety, efficiency, and in-flight entertainment, while leg room has decreased. Travel times, however, have plateaued. When I was ten years old, my family took a trip, flying from New York to LA in about six hours. Forty-five years later, on my most recent trip to the West Coast, the same trip took about…six hours. That is not something my ten-year-old self would have predicted.

This commercial airline speed limit is partly due to the sound barrier, 761.2 mph or 1,225 km/h (called "Mach 1"). Going faster than sound in air causes a sonic boom (a shock wave in the air that travels with the jet producing a trail of explosive sound), and a host of engineering challenges. We can do it, but it has been easier for airlines to stay below this limit. In fact, it is difficult to approach this speed, as parts of the airframe may have air moving across it at faster than Mach 1, causing incredible stress. The speed at which this happens is called the "critical Mach number."

Airlines are also highly motivated to make each flight as cost-effective as possible. Airspeed is part of this calculation. As jets go faster, air resistance increases fuel use. Drag is about proportional to the square of velocity. This means that a 10 percent increase in speed results in a 21 percent increase in fuel use. So, airlines mostly stay around a safe, comfortable, and cost-effective Mach 0.85, and have been for the last half century.

Will we be stuck with this speed limit for the next century or so? Perhaps. The only other option is to break through the sound barrier with supersonic craft. You may remember the Concorde commercial supersonic passenger liner that operated from 1976 to 2003. The plane flew at just over Mach 2 and could get from New York to London in under three hours. However, tickets were several times higher than a typical commercial jet flight. This may have contributed to the service ending—it could not survive through a downturn in the industry after a crash of a Concorde in 2000, and after the September 11, 2001, terrorist attacks. Other factors were involved as well, such as the inability to travel to overpopulated areas, limited range, and competition from more luxurious first-class options.

Two decades later, there are various efforts to revive a supersonic passenger jet service, with an eye toward greater cost-effectiveness. Obviously, the technology exists but it's all about whether there will be a market for supersonic travel at the available cost. Exertions to limit the intensity of the sonic boom through improved wing design are also critical and may open up travel routes to supersonic jets.

Beyond supersonic travel there is hypersonic. This simply refers to very fast planes, traveling from Mach 3 to Mach 16. Such velocity requires a redesign of the jet engine to both produce and handle these speeds. One candidate technology being actively researched is a rotating detonation propulsion system, which uses small, controlled detonations in the engines to provide thrust. These engines are more efficient, and so could carry less fuel. Optimistic projections hold that such a plane could fly from New York to LA in thirty minutes.

The only other way to achieve such travel times is with the other potentially disruptive technology for long-distance travel: rockets. Some futurists imagined that earthbound rocket travel might one day replace jet travel. In the 1962 book *The Man in the High Castle*, for example, writer Philip K. Dick invented a future in which the Nazis won World War II and used their advanced rocketry for intercity travel. As we now know, commercial rocket travel never became a reality.

Rockets are hardly new technology, and they were considered for possible long-distance travel even back in the 1950s. The question is—can rocket travel become safe and cost-effective enough to be used for routine passenger travel?

Elon Musk hopes so. He is developing his Starship rocket to not only get people to Mars and to use as a lunar lander (after being awarded a NASA contract in 2021), but also as a form of long-distance travel on Earth. Using this approach, you could go between any two cities in under an hour (take that, Fogg).

The Past and Future of Trains

Like cars and jets, trains are another core transportation technology likely to persist into the future. Even though train travel emerged relatively recently in the eighteenth century, the technology has deep roots. The oldest evidence of using a primitive track for transportation is the Post Track, a prehistoric causeway in the valley of the River Brue in the Somerset Levels, England, which dates to 3838 BCE. The tracks were grooves in limestone, and wheeled carts were pulled by people or animals.

Wooden rails were introduced in 1515 with Reisszug, a track in Austria. In the 1760s, the Coalbrookdale Company in England began affixing iron to the wooden rails to strengthen them. But the real railroad breakthrough came with the introduction of steam power. At the time, England was having a wood shortage crisis. There simply weren't enough trees to fill the needs of their vast navy and heating homes in the winter. Coal, an efficient, if dirty, replacement, quickly became popular.

However, coal mines have a limitation—if you dig too deep,

the mines start to flood. The steam engine was partly developed in 1698 to help pump water out of the coal mines when Thomas Savery created the "miner's friend." This allowed them to go deeper and greatly increased the supply of the much-needed coal.

In the 1760s, James Watt, among his other steam engine improvements, created the reciprocating steam engine that was able to drive a piston and turn a wheel. This became the basis of steam-powered cars that could carry coal on the existing rails—and the basis of the steam locomotive. In 1812, Matthew Murray developed the first commercially successful steam locomotive for the Middleton Railway in Leeds called the Salamanca.

Over the next two centuries, railways became a major method of transporting both goods and people. Even with the advent of air travel, trains remain popular for medium intercity travel—in Europe, the UK, Asia, and in some parts of the United States like the Eastern corridor where cities are close together. What is the potential future for trains and similar technologies to fill this travel niche?

One somewhat new concept is the hyperloop. Similar to a subway, it is an underground enclosed tunnel that can be evacuated to reduce air pressure and thereby reduce air resistance, making trains faster and more efficient. Long-distance hyperloop trains can theoretically achieve speeds of 600 mph (965 kph) or more. Being underground, they can bypass obstacles and would not disrupt the landscape.

A limiting factor for the deployment of hyperloop systems is the need to build massive long-distance tunnels. For this reason, our old friend Elon Musk founded the Boring Company, in an attempt to drastically reduce the cost per mile of boring tunnels.

Hyperloops also depend on maglev technology. Maglev trains, which are in service in Europe and Asia, where train travel is more popular, use powerful magnets to float trains above the track and to propel them. They are all electric and fairly efficient and can achieve speeds up to 375 mph (600 kph). There is a move to build a maglev in the Eastern corridor of the United States, but there are no concrete plans as of yet. The initial investment is relatively high, and the need to obtain rights of way across hundreds or thousands of properties makes them impractical depending on the local government and regulations. This problem is bypassed using a hyperloop tunnel.

Maglevs and hyperloops will likely have a future, but it's unclear if they will dominate or remain in their current niche. It's possible they will morph from mass transit (cars holding hundreds of passengers) to more individualized transport, using small vehicles with a capacity in the range of four to eight passengers and taking more direct routes to your destination without frequent stops. This, however, would require a more elaborate system of tracks.

As we have seen, there are many possibilities for the future of travel, both personal and mass transit. It seems likely, though, that for the foreseeable future, transportation will continue to be dominated by planes, trains, and automobiles—as we see again and again, old technologies have a remarkable persistence into the future. There is the possibility of some disruption, with supersonic jets, hyperloops, and possibly even rockets, but all are uncertain.

Another way that transportation might be upset is in shifting the distinction among short, medium, long, and intercontinental distance travel. Right now, cars are short to medium distance, trains are mostly medium, and planes are long distance and intercontinental. But what if flying drone cars become so advanced that they can be used for any distance? This could largely obviate the need for other forms of travel.

The real question is, do we have any other Phileas Fogg moments in our future, where the world fundamentally changes because of a disruption in transportation technology? We can extrapolate out current technology and predict that generally everything will get faster, safer, and cheaper, and the world will continue to get effectively smaller. But will we see entirely new forms of travel?

We will likely have to wait for sci-fi technologies that don't exist yet (and may never exist). In the 2012 remake of *Total Recall*, a train system dropped through the center of Earth, a ride that took forty-two minutes. This would make it possible to commute to the other side of the planet. Instant teleportation would essentially erase distance, making everything local. Perhaps the next such milestone in transportation will be to travel around the world in 0.80 seconds.

11. Two-Dimensional Materials and the Stuff the Future Will Be Made Of

We may be entering the age not just of advanced materials but of smart materials.

I have long argued that material science is greatly underappreciated. We can, of course, learn new and better ways to manufacture items out of existing materials, but discovering a new material could be a game changer. It opens up entirely fresh possibilities, the potential for new technologies and capabilities.

Look around your environment right now, wherever you happen to be. It is likely that most of the material you are surrounded with is something humans have been using for thousands of years. One notable exception is plastics, a truly modern material. Will this continue to be true into the future? What new futuristic materials will make that future possible?

Most common materials used to build the modern world have deep historical roots. This is not surprising, as nature contains many materials, and it was not difficult to find ones with the exact desirable properties for different purposes.

The oldest materials used by humans for tools and building were wood and stone. Stone is hard, the hardest substance to be found in most environments with the oldest evidence of stone tool use going back to 3.3 million years ago. This makes it likely that prehumans were using unmodified stones prior to that. The Stone Age of technology lasted from then to about 5,500 years ago.

Despite its age, the global natural stone market in 2018 was estimated at $35 billion and is continuing to grow. Stone remains popular as a building material and for many applications, like countertops, because it is still the best option in terms of aesthetics and quality. Wood also remains in widespread use for its beauty and its physical characteristics. In developed nations, more than 90 percent of homes are still built from wood. Most of the rest are framed with concrete.

Concrete was first used by Nabataea traders in what are now Syria and Jordan 8,500 years ago. It became a dominant construction material in Rome starting around 200 BCE. They made it out of a combination of ash, lime, and seawater. Concrete is currently the single most common building material used today, with about 2 billion tons produced each year and continuing to grow.

Glass was discovered about 4,000 years ago. Ceramics go back even farther, to about 30,000 years ago. Paper is 2,000 years old. The oldest leather artifact is 5,500 years old, but it's likely that use of hide and leather goes back much farther. The oldest known form of cloth is linen dating back to Egypt 5,000 years ago.

The ability to smelt, cast, and forge metals was a transformative technology for human civilization, so much so that archaeologists use them to define technological ages. The oldest metal artifact is a copper awl found in the Middle East from 8,000 years ago. Bronze, an alloy of copper and tin, came soon after, 6,500 years ago. Iron, the relative newcomer, was found in artifacts from Egypt 5,200 years ago.

The Iron Age really took off, however, with the discovery of steel, an alloy of iron and carbon, about 4,000 years ago. Since then, metallurgy has been a critical material technology for human civilization and has continued to steadily progress. According to the World Steel Association, 1,808 million tons of steel was produced worldwide in 2018. There are 20 elements commonly combined with iron (in addition to carbon) to make about 3,500 different grades of steel. The precise content of carbon and other alloys, and the heat treating of steel to control crystal size, also greatly impacts its properties.

One dramatic example of how important steel continues to be in our modern world is the fact that when SpaceX was designing their Starship, which they hope to send to Mars, they decided to build it out of stainless steel. Steel is therefore still a "space age" material.

Material science is perhaps the most extreme example of the futurism principle that old technologies tend to persist much longer into the future than we might naively imagine. Our civilization is still built mostly out of ancient materials that we have been using for thousands of years.

Of course, there are also some important industrial-age materials that have been added to the mix, like rubber and graphite. Some modern materials are simply recently discovered elements, such as tungsten, a metal with a high melting point, central to making tool steel. Uranium and other radioactive materials are necessary for fission and nuclear-based technology. Rare earth elements, such as cerium, neodymium, and terbium, are important to many modern electronics—anything with a battery or magnet likely has them.

One industrial-age material that is critical to our modern world but you probably take for granted is aluminum. It was discovered in 1825 by Danish chemist Hans Christian Oersted, but the process to purify it was costly. For this reason, and because of its unusual properties—it's very light for its strength and heats up and cools down very quickly (has a low specific heat)—aluminum was extremely valuable. It was more expensive than gold for part of the nineteenth century.

However, aluminum is the third most common element in Earth's crust at 8.3 percent (after oxygen and silicon). It's literally in our dirt. By the end of the nineteenth century, the technology to cheaply purify aluminum was coming together. In 1863, two people almost simultaneously figured out how to mass produce the metal: American Charles Martin Hall, who was only twenty-two, and French chemist Paul Héroult, who was twenty-three. Their discovery led to the mass production of aluminum and the steady drop in price, down to only 20 cents per pound by 1930. Today the price hovers around a dollar a pound.

This is a great example of how easy it is to miss a disruptive technology. Cheaply purifying aluminum from dirt was likely not on many nineteenth-century people's lists of technologies that would transform their future, but this obscure chemical process did just that. Aluminum is so cheap and plentiful that we literally use it as a disposable wrap for food. The metal is also highly recyclable, making it useful for canning and its light weight makes it ideal for many structural purposes. There are about 400 pounds of aluminum in a typical modern car, and this figure is growing. The airframe of many modern jets is about 80 percent aluminum. This is truly a modern material that is likely to continue to be important for a long time to come.

But perhaps the most iconic industrial age/modern material is plastic, a purely synthetic material emerging from the science of chemistry. Celluloid is a combination of nitrocellulose and camphor. It was first made by Alexander Parkes in 1856 but was patented as celluloid by John Wesley Hyatt. It was first created as a replacement for ivory, which was running short because of demand and overhunting of elephants, but the moldable material was soon adopted for a wide range of uses.

Try to imagine our modern life without plastic. There are many types of modern plastics, including polyethylene, PVC, acrylic, nylon, polypropylene, and others. They have a range of desirable properties, including hardness, stability, and the ability to mass produce and shape as desired. Plastic can also be sterile, which is critical for the medical industry.

Of course, there is a downside to the plastic revolution—it is so stable that many forms do not significantly biodegrade. Some plastics shed chemicals into the local environment or break down into smaller and smaller particles. The world produces about 300 million tons of plastic waste per year, which is clogging our oceans, and breaking down into microplastics that are getting into the ecosystem and our food chain. Researchers are now looking for biodegradable and more sustainable substitutes for plastic, and in some cases companies are going back to older natural materials (like paper straws instead of plastic straws) because they are safer for the environment. Another example of how we cannot make simplistic assumptions about the priorities that will shape the future.

In addition to ancient and industrial-age materials, what "space age" or advanced materials are important to our infrastructure and technology today? Mostly modern materials are improved versions of older materials, such as advanced ceramics, modern metallurgy, and treated wood, but there are some high-tech materials worth noting.

Composites include materials made of two or more substances that remain distinct in the final material. The idea is to combine materials with complementary properties, such as strength, flexibility, and hardness, to produce a final product with the best of each. For example, there are two general types of composite-reinforced plastics that have advanced properties—fiberglass and carbon fiber. Fiberglass can be easily shaped into a rigid frame and is used for things like boats. Carbon fiber has a high strength-to-weight ratio and is very rigid, so if you need these properties and price is no object, such as military aircraft, then carbon fiber is the way to go. Other advanced composites include shape memory polymer, high-strain composites, organic matrix/ceramic aggregate composites, and particulate wood composites.

There are an estimated 300,000 different materials used today, but we're here to discuss the cutting-edge materials that currently exist, at least in some form, that are likely to be increasingly important in the future.

The goal of modern material science is to create material with a combination of extreme properties, such as material with high strength or rigidity for its mass (referred to as "specific" whatever—specific strength is strength for a given mass), as well as materials with optimal thermal, optical, and/or electrical properties.

Nanostructured Materials

It is safe to say that we have discovered all the stable elements that exist. The periodic table shows us where the empty spaces are, and they have all been filled in. Newly discovered elements are being added to the end of the table, but they all have an extremely high number of protons and are inherently unstable. There is some theorizing about "islands of stability" at some places at the high end of the table—configurations of neutrons and protons that fit together just so, leading to increased structural stability. But "stability" in this context is relative; these are still not elements we will be building stuff out of.

Chances are that there is no unobtainium, vibranium, adamantium, mithril, beskar, kryptonite, red matter, or other high-tech element out there to be discovered. These fictional materials, therefore, must be considered alloys or allotropes of known elements. If we want advanced materials with desirable physical properties, we will need to make them out of elements on the periodic table, which means creating new alloys (mixing different elements together), allotropes (different molecular configurations of an element, like diamond and graphite are to carbon), or composites (mixed materials that remain discrete). Alternatively, it could mean controlling the internal structure of a material.

The technology that alters the structure of a material to control its properties is thousands of years old. Ancient examples of this include heating clay to harden it, or heat-treating steel to control the crystal size, which affects strength and hardness.

The advanced version of this technology is collectively called "nanostructured material," which involves controlling the structure of materials at the nanoscale, from 1 to 100 nanometers (a nanometer, nm, is one billionth of a meter). To put this into perspective using that universal example of thinness, a human hair is between 80,000 and 100,000 nm thick. So, very small.

"Nanostructured" refers to any material that is made of discrete parts on the nanoscale, or has nanofeatures either internally or on the surface. A nanoparticle is in the nanoscale in all three dimensions (no dimension bigger than 100 nm). A nanotube (hollow) or nanorod (solid) has two of three dimensions on the nanoscale. A nanosheet has one dimension on the nanoscale, and if one of the other two is very different from the third (much longer than it is wide), it is called a "nanoribbon." Further, if a nanosheet is a single molecule thick, we might refer to this as a two-dimensional material (it's not literally two-dimensional, but as close as matter can get).

Additionally, nanomaterials can be engineered, incidental, or naturally occurring. Mostly we will be talking about engineered nanostructured materials.

The most commonly discussed nanomaterial in recent years is carbon nanofibers. Carbon can form four strong bonds, and this allows it to form into many stable allotropes, one of which is a two-dimensional sheet of carbon in a hexagonal lattice (picture chicken wire). This allotrope is called "graphene" and is the thinnest two-dimensional material known. Graphene can be rolled up into carbon nanotubes or stacked to form carbon nanofibers. A football field–sized sheet of graphene would weigh only 3.8 grams.

In addition to being thin, graphene is an incredibly strong material, stronger than steel or Kevlar. It has a higher tensile strength (the amount of force necessary to stretch something until it breaks) than any known material—130 GPa (gigapascals). Graphene is harder than diamond, all while being more flexible than rubber.

Graphene also has high thermal and electrical conductivity. It conducts electricity ten times faster than silicon while using less energy. Electrons move through the material at 1/300 the speed of light. For this reason, it could be the basis of an entirely new electronics and computing technology. This could potentially allow for electronics that are smaller, faster, and use less electricity with more efficient cooling than today's technology.

First isolated in 2004, graphene is obtained simply by sticking tape to graphite—a graphene layer will come off (graphite is essentially a stack of graphene layers). Current applications mostly include using carbon nanorods as part of composite material, to provide strength, stiffness, and light weight to things like tennis rackets. It is also used for heat dissipation in small electronic appliances like light bulbs. You can also 3D print with a graphene composite. These humble uses stand in contrast to the hype about its amazing potential.

The disconnect is due primarily to two factors. The first is that the research into graphene is still in the early stages. Much of this research focuses on doping—putting other elements into the basic graphene lattice in order to tweak its physical properties. Graphene, therefore, actually has the potential to be hundreds of specific materials, each fine-tuned for a specific application.

The main limiting factor for graphene is the ability to mass produce it with sufficient quality. Low-quality graphene, with lots of kinks and breaks, does not perform well. A small break, for example, could cause graphene to "unzip." Try pulling apart an intact sheet of paper. You can't do it. Then make a small tear and ripping the paper becomes very easy. The same is true of graphene.

Graphene is therefore another perfect example of the challenges of futurism. On the one hand, it has amazing properties that make it potentially a revolutionary material. Graphene might be for the twenty-first century what plastic was for the twentieth century, and more. On the other hand, if we cannot figure out how to get past the manufacturing hurdles, its ultimate utility might be significantly curtailed, limited to high-end applications but never mass production.

That said, it seems very likely that the graphene family of materials will have a prominent role to play in future technology, but the details are still a bit fuzzy. The critical factor will be the ability to mass produce types of graphene with very few errors. We should also keep our eye on other 2D materials, like boron hydride, which may eclipse graphene in some applications. Molybdenum disulfide is proving to be a promising light source, producing light when heated that is ten times brighter than the 3D version of the material.

Two-dimensional materials are not the only potential nanostructured materials of the future. Foam metals are another emerging technology. These are metal alloys that are engineered to have small spaces in their interior structure. The pores don't have to be nanosized, but they can be and are then called "nanoporous foam metals." These gas-filled pores can be connected or isolated. The result is a material that can be very stiff and strong, but much lighter than the solid metal. There is typically between 5 and 25 percent metal in the final volume. Having a high specific stiffness or specific strength is ideal for many applications, including just about any vehicle, from cars to commercial jets to rockets, as well as lightweight bulletproof armor.

The foam structure could also be insulating against radiation. This will be critical for the future of space travel, but blocking radiation is also useful to contain fission reactors or nuclear waste, or for shielding in medical uses of radiation such as X-rays or radiation therapy.

Additionally, nanostructuring materials can be used to improve the properties of existing materials. This is a big area of research in battery technology—nanostructure the cathodes and anodes, for example, to improve battery characteristics. There is also research into nanopore batteries, using a nanopore foam substance with each pore containing a tiny battery.

Research into nanostructured materials is going in so many directions, it is impossible to know which materials specifically will turn out to be useful. In the aggregate, it seems clear that they will be important high-tech materials for the next century.

Metamaterials

Metamaterials are a form of engineered nanostructured materials that deserve their own explanation. The term "meta" means the material has properties that do not occur in nature. Their properties derive from their nanoscale structure, not from the chemical or physical properties of the elements they are made from.

Such materials include either surface or internal structure composed of nanoscale composite units that can be arranged geometrically to tune their physical properties. This gives an unprecedented level of control, even to the point of creating properties that are not reproduced in any natural material.

For example, a metamaterial may have a nanostructure with features smaller than the wavelengths of light with which they interact. This can allow for some bizarre physical properties, such as having a zero or negative refractive index of light. The refractive index refers to how much a transparent material, like glass, will bend or refract light. With a positive refractive index, light bends in the direction of propagation (like looking at a stick poking into water that appears to bend at the water surface). A negative refractive index means the light is bent in the opposite direction and can also result in reverse propagation of electromagnetic waves.

Since the refractive index is dependent on the geometric arrangement of the nanostructures, this property can be tuned as desired. It is not dependent on the material itself. What are the potential applications for these optical metamaterials? One is making tiny lenses for cameras, lenses that can magnify and zoom. While small cameras for things like cell phones have improved incredibly, mostly because of software upgrades, they do not have any optical zoom. This requires multiple lenses too large for a phone but could prove possible with smaller metamaterial lenses.

Optical metamaterials also allow instruments that can focus below the optical limit. Generally, you cannot image or focus on an object that is smaller than the wavelength of the light you are using. This is like pixels in a digital image—you cannot see details smaller than the pixels themselves. This resolution limit (technically the diffraction limit) has long been considered absolute, but metamaterials have essentially accomplished what was thought to be impossible. While the kinks are still being worked out, this technology appears to be on the cusp of transforming photography, sensors, displays, and microscopy.

Another application, known as "optical cloaking" and often hyped in the press as duplicating the effect of Harry Potter's "invisibility cloak," is the ability for optical metamaterials to become invisible to specific wavelengths of light. While this is not quite invisibility, it can have applications for high-tech camouflage.

Aside from visual applications, metamaterials can also have special electrical or magnetic properties. This could allow for a level of control over electromagnetism not normally possible, with implications for all sorts of electrical equipment. The precise implications are hard to predict, like trying to predict in 1800 electrical devices of the future. The broad brushstrokes are electronic devices that are smaller, more efficient, and more powerful. Metamaterials are redefining what is possible when it comes to controlling and manipulating all forms of electromagnetism.

Since we are still at the dawn of metamaterials, there are many paths down which this technology can go. There are experimental possibilities such as extreme energy absorption, or deflecting seismic waves, or even active metamaterials that adapt to their local conditions. Imagine a firefighter with a metamaterial uniform that allows them to stroll unscathed through a blazing inferno.

Smart Materials

Smart materials blur the line between material science and machine technology. The core property of a smart material is that it can change its form or properties based on external stimuli. Basic examples include piezoelectric materials, which generate electricity when a compression force is applied to them.

Shape memory materials can change their configuration, usually between two different states, under different environmental conditions, such as ambient heat. They have an original shape that can then change dramatically when a stimulus is applied, and then return to their original shape when the stimulus is removed. Or you might bend them with physical force, then return them to their original shape by applying heat. When cleverly designed, a flat-shaped memory material can unfold like origami into an elaborate three-dimensional shape.

Applications of this property might include the ability to ship three-dimensional objects in a flat and stackable form, then have them open into their final shape at their destination, without requiring assembly. The property can also be exploited in machines that need to react to changes in their environment, such as responding to a fire by releasing a fire-suppressant gas.

There are also materials that have what is called "pseudoelasticity" or "superelasticity." These materials return to their original shape automatically, without the need to apply heat or another stimulus. This is more similar to simple elasticity, such as rubber. The difference is that materials that are not normally elastic, such as metals, are engineered at the nanoscale to display elasticity. Bend an aluminum rod and it will stay bent, but a smart alloy can return to its original shape. Such materials could have important biomedical applications—living things tend to be squishy and flexible, but medical implants are often hard and rigid. With superelastic metals, you could have the needed strength and other properties, with the flexibility required by a living, moving recipient.

Another kind of smart material is chromoactive material. If you are of a certain age, you may remember mood rings, which would allegedly change color with your mood. In reality, they were chromoactive material that would change color with temperature. Other materials can change color based on pressure, light, or other stimuli. Obvious applications here would be indicators or warning signs that become visible in certain conditions, such as overheating or the presence of radiation. As optical electronics are developed (using light instead of electrons to make computers and other equipment), chromoactive materials could be critical to their function too.

Finally, there are magnetorheological materials that can change their properties based on the application of a magnetic field. Magnetic fields are useful because they can penetrate solid objects, and so this property can be used to control devices that are otherwise inaccessible (e.g., because they are inside a living organism).

Smart matter such as these allows for control of materials in multiple ways after they are manufactured, granting a crude type of programmable matter. As this technology advances, the range and precision of control is likely to improve. Small electrical and magnetic fields can easily be controlled by computers, allowing for a connection between software and physical material—changing the properties of matter at will or under the guidance of an AI.

The ultimate smart material, however, is fully programmable matter, using nanomachines or even smaller, that can change their shape and properties almost without limit under the control of software. Imagine building a home entirely by taking a pile of ooze (nanomachines) and then controlling them to form into the digital design on your computer, which you view with AR (augmented reality) goggles as you walk through your home as it is coming into being. The home can trap heat in the winter and ventilate in the summer. The outer surface could be entirely made of high-efficiency photovoltaic material. The house itself could be entirely programmable, reconfigurable at will. You could change colors, move walls, grow furniture, and add windows or lighting fixtures.

This is clearly a far future vision of material science, the ultimate expression of the ideal material. Eventually there may be only one material, and the border between software and hardware will not just be blurred but obliterated. The physical world will become a virtual world, in a way—digital and programmable.

In the meantime, material science will continue to advance, and at an increasing rate. There is every reason to expect that ancient materials—wood, stone, metal, glass, and concrete—will continue to dominate long into the future. The technology to work them will also continue to improve, and newer advanced materials will increasingly be in the mix. Nanorods can be added to concrete to make it stronger and more resilient than ever, for example. Nanostructured materials will increasingly dominate our technology and make their way into construction as well.

Material science itself is being revolutionized by digital technology and AI. We no longer need to depend on trial and error alone to discover materials with superior properties. They can be simulated and designed, with artificial intelligence sifting through millions of possibilities to find optimal mixes and configurations. We may be entering the age not just of advanced materials, but also optimized materials, with the most extreme properties theoretically possible. Imagine a metal with a specific strength that is the greatest possible within the laws of physics.

New materials expand the possibilities of our technology and future ones, like a space elevator, that are not possible with today's materials. We will have to invent them before such futuristic visions can become a reality.

12. Virtual/Augmented/Mixed Reality

Welcome to the new digital reality.

Ready Player One, a 2018 film based on the book by the same name, provides a compelling vision of the near future (2045) when virtual reality has matured and dominates our lives. The planet has descended into harsh economic depression, and most of humanity spends the bulk of their time in OASIS, a virtual world in which they can be anyone, go anywhere, and do anything.

How likely is this vision, in terms of the technology and the effect on society? The film explores some interesting issues. For example, the economy of OASIS, even though it is, essentially, a game, dwarfs any other economy in the world. Many people turn their time and energy into their virtual lives, neglecting the physical world with obvious negative consequences. The virtual world also becomes a convenient mechanism by which the powers that be can control the masses—or at least distract them.

The film also shows how one's persona in the game can be nothing like their physical self out in "meatspace." A disabled middle-aged man could be a dangerous and intimidating presence in OASIS. Age, gender, race, and to some extent even resources mean nothing there, where you can become whatever you can imagine.

What Is Virtual Reality?

The basic concept of virtual reality is to use technology to present stimuli to one or more senses to create a virtual experience either instead of or in addition to the real physical world. Complete immersion is referred to as "virtual reality" (VR), while adding to external reality is called "augmented reality" (AR).

Using an expansive definition, perhaps the first manifestation of VR was the stereoscope created in 1840 by Sir Charles Wheatstone. Two years earlier, he had described the principle of stereopsis—if you present two slightly different images to each eye, you can create the illusion of 3D. You may remember this effect if you owned a View-Master as a child. The toy was introduced in 1939 and features a wheel of pictures you can advance by pulling a small lever as you hold the device up to your eyes. To complete the circle, View-Master is now working on a View-Master Virtual Reality version.

While a stereoscope is primitive, it does represent the basic idea of VR—that you can hack the senses to create a compelling illusion of reality. This relates to concepts discussed in chapter 6 on the brain-machine interface about how our brains construct our perception of reality. This construction has rules, and those rules can be exploited to generate compelling illusions, even illusions that your brain will accept as genuine.

I had a visceral personal lesson in this fact after buying my first VR headset, which came with access to a demonstration game called *Richie's Plank Experience*. While in the comfort and safety of my home office, the VR headset and headphones provided the sensation of walking out on a plank some thirty stories up a high-rise. Of course, I knew consciously that I was just playing a video game, wearing a headset, and that at no point was I at risk of suddenly converting my potential energy into kinetic energy and falling to my death. Even so, the more subconscious parts of my brain completely bought the audiovisual illusion. Utterly convinced, my lizard brain was screaming as if I were at risk of falling and splatting on the sidewalk below.

I take it as a point of small pride that my primate neocortex won out over my more primitive amygdala, and I was able to completely walk out onto the plank. My pride was greatly enhanced when my brother and coauthor Jay failed this test of higher cortical function—he bailed, ripping the headset off his face, and just saying, "Nope!"

You can find videos on YouTube of newbie VR users playing *Richie's Plank Experience* and bravely diving off the plank, while in reality they are taking a header into their TV set. The point is, the illusion works. When a sufficient amount of your visual field is occupied by the virtual world, your brain accepts that as your reality. Such illusions are also greatly enhanced when more than one sensory modality conspires together to reinforce this virtual reality. For example, when your friend is trying the *Plank Experience*, surreptitiously blow a small fan in their face to simulate the high-altitude wind.

VR refers to the use of this technology to completely enter, at least visually, a virtual world. You are then blind to the real physical world around you (VR software handles this by mapping your physical space and then helpfully showing when you get close to the edge).

The first manifestation of something that can be considered VR was the Sensorama built in 1962 by Morton Heilig. Heilig considered this the "theater of the future" and first presented the idea in a paper in 1955 (again, technology typically has deeper roots than we naively imagine). The Sensorama was a multisensory immersive experience, with wide-angle stereoscopic images, stereo sound, fans, odor emitters, and even a mobile chair. By all reports, the experience was rather compelling. Heilig could not secure funding for the Sensorama, and so the project died—perhaps this is another example of how our present could have been very different were it not for quirky chance.

Other precursor technology to VR included the Telesphere,

a head-mounted viewer also created by Heilig in 1960. The 1961 Headsight was a head-mounted motion-tracking display. This device would move cameras to match your head movements so that you could virtually look around. In 1966, the first flight simulator was created, which was the first practical application of this type of technology. Then in 1968, we get the first computer-based head-mounted display—the *Sword of Damocles*.

The technology continued to develop over the next three decades until the first consumer video game VR headsets were released in the 1990s to very little market penetration. The next-generation VR headsets began releasing in 2016, with many predicting that this time VR was poised to go mainstream.

In 2020, the global AR and VR markets were valued at $12 billion, with a 54 percent rate of increase and a projected market of $72.8 billion by 2024. There is always a chicken and egg problem with this type of technology—you need users to make it worthwhile to invest in hardware development and games, but you need games and good hardware to lure in sufficient users. Such technology is often bootstrapped by early adopters, and we appear to be transitioning through the early adopter phase to the mainstream use phase now.

The current state-of-the art VR commercial headsets have a 110-degree angle of view, which is sufficient for an immersive experience but still gives a slight sense of tunnel vision. Resolution is lagging behind the best computer monitors, although it is good enough not to be a limiting factor. Further, eye-tracking technology allows for the software to increase the resolution only where the user is looking, giving an effectively higher resolution without bogging down the whole system.

Controllers are rapidly evolving and already include general-use hand controllers, but also foot controllers to track leg movements and finger controllers to allow for detailed manipulation. Devices also have what is called "haptic feedback" or tactile sensory feedback. The hand controller, for example, could vibrate to indicate physical resistance when manipulating a virtual object. VR systems can be wired or wireless, allowing for a great deal of freedom of movement. There are also increasingly specialized controllers for specific games, like car wheels, weapons, and rackets.

As the technology progresses, there are plans to increase the angle of view, give higher resolution, improve motion tracking and eye tracking, and offer greater portability.

There are some downsides to the current state of VR technology, some of which result from successful immersion into the virtual world. The biggest limiting factor is motion sickness, which results from the disconnect between what the user's eyes are seeing and what their vestibular system is sensing. This is the part of our body that identifies motion and the direction of gravity. A healthy brain compares the visual input with vestibular input in real time, and a mismatch can cause motion sickness. VR-induced motion sickness can be profound, making it impossible for some individuals to even use the technology.

I have personally experienced this, especially with vertical movement. If my virtual avatar, for example, runs over rough terrain, while my head in physical space is not moving, that will induce instant nausea, ending the VR experience. The primary fix for this problem at present is to limit the ways virtual characters move. Most VR games and other applications now have an option to teleport your character from place to place, rather than moving them through space. You can still walk around your physical VR space, which doesn't pose the same problem because you are moving as well. While an effective fix, this does limit the kinds of experiences available in VR.

Another approach to the motion sickness problem is to provide vestibular feedback to match the visual motion. This technology is in the experimental phase, using, for example, little electrical pulses behind the ear to stimulate the vestibular system. It remains to be seen if such a neurological hack will work.

The ultimate solution would be to match the physical movements of the VR user with that of the virtual avatar. This is the system used in *Ready Player One*—the VR user is on a platform like a running track but capable of moving in any direction. This platform would have to be able to move up and down to match vertical movement, and/or the user can be attached to a harness.

Another limiting factor for VR is the current size and weight of the headsets that need to be strapped to your face. This can cause fatigue and limits the duration users are willing to use VR. This factor has curbed, for example, the adoption of virtual offices. Eight hours is just too long to wear a VR headset.

The solution to this is obviously smaller and lighter headsets. This imperative, however, is running up against the desire for higher video resolution and greater angle of view. Companies will continue to explore the sweet spot compromise as the technology continues to incrementally advance. Eventually we will likely have the perfect headset, such as wraparound glasses or even contact lenses.

Augmented Reality

AR is similar to VR except that it does not replace our ability to see the world; it overlays graphics onto the outside world. The first mainstream introduction to this technology was perhaps the *Pokémon GO* mobile game. In this game you hold up your smartphone or tablet and the camera displays a real-time view of the world overlayed with Pokémon creatures. The game was a sensation, but like all new fads, it rapidly dwindled to a base of enthusiasts.

A more high-tech version of AR is the now infamous Google Glass (or just Glass), first introduced in 2013. These looked like regular glasses, with a small camera incorporated. The glasses were also a heads-up display for your smartphone, which could be controlled with voice commands. Google very deliberately released an early adopter "Explorer" version for $1,500, a little pricey for a mainstream consumer device.

The real pushback came from the camera, which raised concerns about privacy. It also gave rise to the term "Glasshole" for people who would not take off their Glass device when requested. Google continues to develop Glass, focusing now on their "Enterprise" edition (optimized for business applications rather than the consumer). This allows workers to have access to manuals and checklists on the fly, as well as produce video documentation for quality control.

There are some significant advantages with AR over VR. AR glasses can be much lighter because they do not have to produce a full view of graphics. They can be run by a smartphone, rather than a computer with a high-end graphics card.

Perhaps more significant is the lack of any motion sickness. Since you still see and move around in the real world, there is no disconnect between vision and vestibular sensing. This also makes AR more portable and mobile—you are not tied to a VR physical space.

There are still concerns with AR. As mentioned, there are privacy issues with wearing a device that is designed to record whatever you see. The flow of AR information also goes both ways—wearing an AR device gives someone access to information on the sly. Imagine, you could be looking up personal information about someone as you talk to them.

This is an aspect of AR that makes specific applications and degree of adoption difficult to predict. It may be like using a cell phone with in-ear devices. At first it was odd to see someone walking down the sidewalk apparently talking to themselves, but eventually we got used to the sight of someone talking on their phone in this way.

As we've also become terribly familiar with, it can be rude and intrusive to constantly be text messaging during dinner or checking your email or social media. Imagine if this can be done through your glasses. Are you looking at me or at the latest stock prices? Will this be a deal killer for many AR apps, or just become another part of life?

The final concern with AR is safety. There is already a problem with distracted drivers and even pedestrians due to cell phones and other portable devices. People walk out into traffic or fall into open trapdoors while texting their friend. Imagine how much more distracted people would be if they have constant information to focus on in a heads-up display. AR may ultimately be safer in some situations than looking down at a smartphone, but research will be necessary to sort out the safest behaviors as AR becomes more popular.

For completeness within these pages, I must note that there is also something known as "mixed reality," a term introduced in a 1994 paper by Paul Milgram and Fumio Kishino called "A Taxonomy of Mixed Reality Visual Displays." Mixed reality blends the digital and physical worlds in a continuum between VR and AR.

There are mixed-reality arcade rooms that have physical props and walls to match the virtual world. Anyone who has leaned up against a virtual wall in a VR game only to fall over knows how much more immersive it could be if those virtual walls were real. Having digital holograms of people or objects in the physical space is another example of mixed reality. This may be the dominant form of virtual technology in the future because it blends the best of the physical and virtual worlds into a seamless experience.

VR and AR Applications and the Future

The technology for VR and AR is here and is being used. It continues to advance, and more applications are becoming available. Where will this lead?

In the near term, we can expect incremental developments in both hardware and software that will continue to improve the VR/AR experience. This includes cameras that can track a user's movements, making the need to manipulate a controller obsolete. As you move around, your avatar will exactly match your movements. Controllers would be optional add-ons for specific functionality, allowing you to interact physically with virtual objects, like a steering wheel or golf club.

Haptic feedback is also becoming more sophisticated, but for now limited to vibration or gyroscopes to produce resistance. As this tech advances, we will likely see in a few decades something like the VR suit in *Ready Player One*—a whole body outfit that translates virtual touching into real tactile sensations. The goal is to render anything that happens in the virtual world into real physical sensations.

With current technology, the most common application for VR is gaming, which is driving the technology as it drove the development of graphics technology in general. VR games are already incredibly fun and immersive, but not every type of game is suited to VR. As with many technologies, the new does not always completely replace the old. For a long time we will probably continue to see traditional games played on a PC, console, or handheld device alongside games developed specifically for VR.

VR will expand to other forms of entertainment. VR movies already exist mostly for demonstration purposes. I watched one and, while it was interesting, it was difficult at times to determine where I was supposed to look. A great film is partly about managing the experience of the viewer, including directing their attention. This is more challenging in a 360-degree environment. The question becomes whether VR movies will evolve as an art form, with artists figuring out how to take advantage of the medium, or whether it will remain a niche curiosity.

VR education again will likely become an add-on but not

replace older methods. A high-resolution 3D scan of a museum, archaeological site, lunar or martian surface, geological formation, or other site of interest can be viewed in VR and provide an excellent learning experience. There are obvious advantages here in terms of expense, accessibility, and being able to layer in educational information.

VR educational experiences are not limited to places but also can be applied to events and objects. Imagine learning about the Battle of Gettysburg by living through it as not just one soldier, but many different soldiers with different perspectives. There is already a VR experience that allows you to travel on the Apollo 11 mission as an astronaut, in real time.

Using VR, we can also examine a fossil or artifact. More profoundly, this would not be limited to typical human perspective. We would be able to examine a molecule, a microscopic organism, or at the other end of the spectrum, galaxy cluster formations. Much of this can be accomplished with traditional video, but VR gives you the opportunity to experience it in 3D, to walk around and take different perspectives, manipulate the object, and have a much more immersive understanding.

As VR advances, it will also become more useful for business applications, such as meetings, seminars, and even day-to-day office work. A virtual workspace has tremendous advantages, making it easier to collaborate with others displaced in both space and time. Combined with a virtual presence, VR could dramatically reduce the need and utility of traveling to a location to do work or meet with others, with tremendous gains in efficiency. Sitting in a comfortable chair in your home office, a robust VR setup could connect us to anyone, anywhere, to any online information, and to countless apps for work, communication, and recreation. Once sufficiently portable and wireless, you could even go about your day at home while working virtually.

But again—we cannot predict that people will want to do this simply because the technology advances to the point where they can. Living in a virtual world may become tiring and cause people to pine for the raw genuineness of primitive physical contact. The COVID pandemic starting in late 2019 highlighted the limits and frustrations of replacing in-person experiences with purely virtual ones. However, it also demonstrated some potential advantages, such as the rapid acceptance of telehealth visits to augment in-person doctor visits. The psychological, cultural, and societal implications of widespread VR adoption could be extreme.

I suspect that VR will be increasingly adopted, but not necessarily replace more traditional methods in any sphere. The advantages and efficiencies are too great to ignore, but history has shown that simple traditional methods have an enduring appeal.

It is also not our current selves that will exist in this VR future. The people of the future will likely be far more adapted to VR than we are. Just as the current younger generation quickly became more adjusted to ubiquitous social media than their parents, they may similarly struggle to understand their kid's easy adoption of VR.

Interestingly, AR may have more acceptance and a more profound impact on our lives than VR. AR has the huge advantage of being usable while still existing in the physical world. Imagine an AR GPS system that doesn't just tell you where to turn but that shows you an obvious path to follow. You never have to take your eyes off the road. AR GPS could also highlight your destination and give you useful information about it. While traveling in unfamiliar territory, an AR system could overlay a wealth of data onto the streets and buildings, refined and filtered through your simple verbal commands. Where is the best place to park? What are the best restaurants in the area?

We could also envisage shopping with an AR assistant, helpfully providing you real-time information about any prospective purchase. This could give reviews, places to purchase the item more cheaply, or a warning if the item is a knockoff or a fraud.

AR could also be your secret social advisor, reminding you of the names of people you casually know and basic—or even critically important—facts about them. Or it could serve as an incredibly powerful expert assistant, overlaying the schematics of a machine for an engineer to follow, for example, or the medical scans for a surgeon, even highlighting critical bits of anatomy to avoid, or cancerous tissue to remove. In TV series *The Book of Boba Fett*, this was demonstrated when a droid aided in the reconstruction of a ship by using a hologram to show exactly where a piece was supposed to go.

In combat situations, AR could help locate and identify enemy soldiers, while also highlighting allies and civilians. They can help soldiers avoid traps and explosive devices.

When traveling in a foreign country, an AR assistant could provide real-time translation services and currency exchange information.

AR gaming may also be superior to VR gaming in some ways. You could play in your backyard, in a park or natural setting, or even in designated AR gaming areas. This allows for a large gaming space and the ability to move around freely, while the space is filled with zombies or aliens to shoot (although it may be a safety issue having kids with toy guns running around), virtual items to find, puzzles or mysteries to solve, or people to interact with.

AR could also be used to overlay, or skin, reality with any cosmetic changes desired. Holiday decorations on houses may be purely virtual, making an incredible haunted house for Halloween that can click over to a winter wonderland for Christmas. AR parties could also involve not only skinning the environment, but everyone present could also be augmented to look like anything they want.

Advanced AR technology could essentially overlay a completely different reality onto the physical world, or enhance it, while providing you a stream of useful information. It's possible we could spend more of our time experiencing the world with AR than without.

The Metaverse

In 2021, Facebook CEO Mark Zuckerberg revealed his plans to develop what he calls "the metaverse," a term coined in the 1992 science fiction novel *Snow Crash* by Neal Stephenson. Think of the metaverse like the World Wide Web, a platform that exists on the internet, but in mixed reality (VR, AR, standard computer interface, and physical reality). He envisions a decentralized network of applications, much like the internet itself, not controlled by any one company.

It remains to be seen if this will be the killer VR app that fully transitions mixed reality into the mainstream, or if the hardware technology is simply not ready for this level of use. The fact that a company with the assets of Facebook is going all in on the metaverse, however, could be critical. Zuckerberg could solidify his place as the Ford of social media. If the metaverse survives (and again, being backed by a multibillion-dollar company is a positive predictive factor), it could trigger a technological feedback loop, where increased demand motivates investment in improved hardware technology, which in turn drives more demand.

Even if the metaverse becomes the primary way we access information and computer applications in the near future, it will likely not replace existing methods but will augment them. We'll still be shopping, cruising social media, and communicating on our smartphones, but in mixed reality.

Neural Reality

There is one more form of virtual reality, one that represents the ultimate expression of this technology, neural reality (NR). With NR, you don't have to strap on goggles or glasses or deal with issues of safety or a disconnect between the physical and virtual worlds. NR is essentially the Matrix—using a brain-machine interface to directly feed the virtual world into your mind. When this technology is perfected, NR would be absolutely seamless. Short of perfection, it could still be a compelling interface.

NR is necessarily not mixed reality but rather completely replaces your sensory input and motor output with a virtual world, while you, for example, lie safely in bed. There is no reason to think that this technology is not possible; it is only a question of how long it will take and how good the technology will be.

How will individuals and society react? All the applications I discussed above, from entertainment to training to education, would be possible with NR but better and more compelling. There's no question it will be seductive to live in a virtual world where you can have literal godlike power.

Alternatively, you could live any life, die hundreds of times without consequence, experience any time and place, or completely change yourself. Some futurists have even speculated that some form of neural reality might be the ultimate destination for most technological civilizations. This may be the answer to the Fermi paradox—where are all the aliens? Living in their own virtual worlds. Perhaps we'll let robots and AI run the physical world while we live as gods in our virtual ones. Perhaps this fate will be available only to the rich and the elite while most humans must slum it in the physical world.

The implications for human civilization are profound. Perhaps use of NR will be strictly limited, to keep people physically fit and productive (with exceptions granted for those with physical disabilities). There could also be issues of people losing the ability to tell if they are in the real world or not, with some not afraid of dying because they falsely think they are in the virtual world. Likely there will be new mental disorders deriving from excess NR use: NR overuse dysphoria, reality maladaptation, NR-reality confusion, and so on.

The spectrum from AR to VR and NR is all extremely powerful

technology. Like all powerful tech, there is tremendous potential for use and abuse. The most likely future is one in which everything happens to some extent. The individual and collective choices we make will determine if the future is an augmented reality utopia or a world of NR zombies looked over by pitying AI robots.

13. Wearable Technology

Imagine wearing a suit that fits like a second skin and would be every tool ever made.

There are few things that separate humans from the rest of the animal kingdom. Some animals can use tools, communicate, punish their members who break social norms, or recognize themselves in the mirror. One feature that stands out, however, is that humans wear things to enhance their abilities and adaptability. (If wearable technology is a marker for advancement, then perhaps this makes Inspector Gadget the pinnacle of human development.)

A few animals, such as certain crabs and insects, could be said to "wear" things for protection or camouflage, but mostly animals live their lives naked. It is difficult to tell for sure when humans started wearing clothing, but paleontologists estimate this was around 170,000 years ago, based on the genetic divergence of lice that inhabit clothing. Our closest cousins, Neanderthals, also wore clothing.

The first "wearable technology" was likely a simple fur cape, but it allowed humans to adapt to colder climates without having to evolve fur. Eventually this developed into tighter-fitting stitched clothing. From these humble beginnings, what is the state of the art of wearable technology, and where is it likely to lead?

Clothing evolved beyond mere temperature regulation. It also serves as protection, for adornment, humility, and to signify status and membership in a specific segment of society. One of the earliest and greatest innovations in wearable technology was the pouch. The oldest example known comes from the Iceman, Ötzi, who lived between 3400 and 3100 BCE. To his belt was attached a leather pouch that contained three flint tools, one bone awl, and a lump of *Fomes fomentarius* (tinder fungus). We take this for granted, but the ability to carry around important tools while freeing up the hands was the pinnacle of innovation at the time. Pockets were not invented until the 1600s, so pouches remained cutting-edge wearable technology for thousands of years.

Another early example of wearable technology is eyeglasses. They were invented in northern Italy, most likely in the town of Pisa. In a sermon delivered on February 23, 1306, the Dominican friar Giordano da Pisa (c. 1255–1311) wrote, "It is not yet twenty years since there was found the art of making eyeglasses, which make for good vision." Again, we take something simple like eyeglasses for granted, but that was an amazing advance for those with refractive vision problems. Eyeglasses (or some version of them) were also used to enhance normal vision, functioning as a magnifying glass or a filter to reduce glare, such as sunglasses.

Next, we move to Nuremberg, Germany, in 1510 where Peter Henlein invented the pocket watch. Actually, his version was worn around the neck and was shaped like an orb (remember, pockets weren't invented yet). Once waistcoats with pockets became fashionable in the 1600s, the more recognizable pocket watch was developed. So now wearable technology extended to devices that provide information. This tech advanced in 1904, when the aviator Alberto Santos-Dumont pioneered the use of the wristwatch, which allowed him to have his hands free when piloting.

People kept developing more places to wear more kinds of impressive gadgets. Wearables can now serve not only as clothing, but also as protection (armor, work gloves, hard hats, goggles), to enhance or correct our senses, enhance strength and mobility, free our hands from having to carry specific tools or gadgets, carry our belongings, act as portable or concealable weapons, provide us conveniently with information, or monitor our own biological functions or activities. There is wearable tech for specific environments and situations, like scuba gear for diving, parachutes for jumping out of planes, and space suits for, well, being in space.

Perhaps my favorite quirky wearable technology is the chair pants (which you may remember from the TV show *Silicon Valley*). This is a collapsible stool attached to the back of your pants that deploys when you get into a sitting position, so you can instantly sit anywhere. They demonstrate (similar to the "fanny pack") that wearable technology often presents a dilemma between fashion and function. Yeah, that's useful, but do I want to look like an idiot wearing it?

There is also an entire category of wearable medical devices. This would include orthotics—braces for bad joints to compensate for weakness or to protect an injury. Hearing aids are another type of common medical wearable.

Many jobs have specific wearable gear and outfits—headlamps for miners and others who need hands-free directable and portable light, welding goggles, wearable magnifying glasses for fine work, protection for firefighters, and a host of military gear and sporting gear.

Wearable tech is everywhere. It acts as an extension and augmentation of ourselves and may even become part of our persona, and it is indispensable for many jobs and activities.

Technology that Enables Wearables

As the name implies, wearable technology is simply technology designed to be worn, so it will advance as technology in general advances. For example, as timekeeping technology progressed, so did the wristwatch, leading to the smartwatches of today. There are certain advances that lend themselves particularly to wearable technology. One such development is miniaturization.

The ability to make technology smaller is a general trend that benefits wearables by extending the number of technologies that are small enough to be conveniently and comfortably worn. We are all familiar by now with the incredible miniaturization in the electronics industry, and especially in computer chip technology. Postage-stamp-sized chips are now more powerful than computers that would have filled entire rooms in prior decades.

As is evidenced by the high-quality cameras on a typical smartphone, optical technology has already significantly miniaturized. There is ongoing research into tinier optics still, using metamaterials to produce telephoto and zoom lenses without the need for bulky glass.

"Nanotechnology" is now a collective buzzword for machines that are built at the microscopic scale (although technically it is much smaller still), and of course, nanotech will have incredible implications for wearables.

We are also at the dawn of flexible electronics, also called "flex circuits" and more collectively "flex tech." This involves printing circuits onto a flexible plastic substrate, allowing for softer technology that moves as we move. Flexible technology can more easily be incorporated into clothing, even woven into its fabric. The advent of two-dimensional materials, like carbon nanotubes, which can form the basis of electronics and circuits, are also highly flexible. Organic circuits are yet another technology that allows for the circuits to be made of flexible material, rather than just printed on flexible material.

Circuits can also be directly printed onto the skin, as a tattoo, using conductive inks that can act as sensors. One company, Tech Tats, already offers one such tattoo for medical monitoring purposes. The ink is printed in the upper layers of the skin, so they are not permanent. They can monitor things like heart rate and communicate this information wirelessly to a smartphone.

Wearable electronics have to be powered. Small watch batteries already exist, but they have finite energy. Luckily there are a host of technologies being developed that can harvest small amounts of energy from the environment to power wearables (in addition to implantable devices and other small electronics). Perhaps the earliest example of this was the self-winding watch, the first evidence of which comes from 1776. Swiss watchmaker Abraham-Louis Perrelet developed a pocket watch with a pendulum that would wind the watch from the movement of normal walking. Reportedly it took about fifteen minutes of walking to be fully wound.

There are also ways to generate electric power that are not just mechanical power. Four types of ambient energy exist in the environment—mechanical, thermal, radiant (e.g., sunlight), and chemical. Piezoelectric technology, for example, converts applied mechanical strain into electrical current. The mechanical force can come from the impact of your foot hitting the ground, or just from moving your limbs or even breathing. Quartz and bone are piezoelectric materials, but it can also be manufactured as barium titanate and lead zirconate titanate. Electrostatic and electromagnetic devices harvest mechanical energy in the form of vibrations.

There are thermoelectric generators that can produce electricity from differences in temperature. As humans are warm-blooded mammals, a significant amount of electricity can be created from the waste heat we constantly shed. There are also thermoelectric generators that are made from flexible material, combining flex tech with energy harvesting. This technology is mostly in the prototype phase right now. For example, in 2021, engineers published the development of a flexible thermoelectric generator made from an aerogel-silicone composite with embedded liquid metal conductors resulting in a flexible that could be worn on the wrist and could generate enough electricity to power a small device.

Ambient radiant energy in the form of sunlight can be converted to electricity through the photoelectric effect. This is the basis of solar panels, but small and flexible solar panels can be incorporated into wearable devices as well.

All of these energy-harvesting technologies can also double

as sensing technology—they can sense heat, light, vibration, or mechanical strain and produce a signal in response. Tiny self-powered sensors can therefore be ubiquitous in our technology.

The Future of Wearable Tech

The technology already exists, or is on the cusp, to have small, flexible, self-powered, and durable electronic devices and sensors, incorporated with wireless technology and advanced miniaturized digital technology. We therefore can convert existing tools and devices into wearable versions, or use them to explore new options for wearable tech. We also can increasingly incorporate digital technology into our clothing, jewelry, and wearable equipment. This means that wearable tech will likely increasingly shift from passive objects to active technology integrated into the rest of our digital lives.

There are some obvious applications here, even though it is difficult to predict what people will find useful versus annoying or simply useless. Smartphones have already become smartwatches, or they can pair together for extended functionality. Google Glass is an early attempt at incorporating computer technology into wearable glasses, and we know how it has been received.

If we extrapolate this technology, one manifestation is that the clothing and gear we already wear can be converted into electronic devices we already use, or they can be enhanced with new functionality that replaces or supports existing devices.

We may, for example, continue to use a smartphone as the hub of our portable electronics. Perhaps that smartphone will be connected not only to wireless earbuds as they are now, but also to a wireless monitor built into glasses, or sensors that monitor health vitals or daily activity. Potentially, the phone could communicate with any device on the planet, so it could automatically contact your doctor's office regarding any concerning changes, or contact emergency services if appropriate.

Portable cameras could also monitor and record the environment, not just for documenting purposes but also to direct people to desired locations or services, or contact the police if a crime or disaster is in progress.

As our appliances increasingly become part of the "internet of things," we too will become part of that internet through what we wear, or what's printed on or implanted beneath our skin. We might, in a very real sense, become part of our home, office, workplace, or car, as one integrated technological whole.

We've mostly been considering day-to-day life, but there will also be wearable tech for special occupations and situations. An extreme version of this is exosuits for industrial or military applications. Think Iron Man, although that level of tech is currently fantasy. There is no portable power source that can match Iron Man's arc reactor, and there doesn't appear to be any place to store the massive amounts of propellant necessary to fly as he does.

More realistic versions of industrial exosuits are already a reality and will only get better. A better sci-fi analogy might be the loader exosuit worn by Ripley in *Aliens*. Powered metal exosuits for construction workers have been in development for decades. The earliest example is the Hardiman, developed by General Electric between 1965 and 1971. That project essentially failed and the Hardiman was never used, but since then development has continued. Applications have mostly been medical, such as helping people with paralysis walk. Industrial uses are still minimal and do not yet include whole-body suits. However, such suits can theoretically greatly enhance the strength of workers, allowing them to carry heavy loads. They could also incorporate tools they would normally use, such as rivet guns and welders.

Military applications for powered exosuits would likely include armor, visual aids such as infrared or night-vision goggles, weapons and targeting systems, and communications. Such exosuits could turn a single soldier into not just enhanced infantry, but also a tank, artillery, communications, medic, and mule for supplies.

Military development might also push technology for built-in emergency medical protocols. A suit could automatically apply pressure to a wound to reduce bleeding. There are already pressure pants that prevent shock by helping to maintain blood pressure. More ambitious tech could automatically inject drugs to counteract chemical warfare, increase blood pressure, reduce pain, or prevent infection. These could be controlled by either onboard AI or remotely by a battlefield medic who is monitoring the soldiers under their watch and taking actions remotely through their suits.

Once this kind of technology matures, it can then trickle down to civilian applications. Someone with life-threatening allergies could carry epinephrine on them to be injected, or they could wear an autoinjector that will dose them as necessary, or be remotely triggered by an emergency medical responder.

Everything discussed so far is an extrapolation from existing technology, and these more mature applications are feasible within fifty years or so. What about the far future? This is likely where nanotechnology comes in. Imagine wearing a nanosuit that fits like a second skin but that is made from programmable and reconfigurable material. It can form any mundane physical object you might need, on command. Essentially, the suit would be every tool ever made.

You could also change your fashion on demand. Go from casual in the morning to business casual for a meeting and then formal for a dinner party without ever changing your clothes. Beyond mere fashion, this could be programmable cosplay—do you want to be a pirate, or a werewolf? More practically, such a nanoskin could be well ventilated when it's warm and then puff out for good insulation when it's cold. In fact, it could automatically adjust your skin temperature for maximal comfort.

Such material can be soft and comfortable, but bunch up and become hard when it encounters force, essentially functioning as highly effective armor. If you are injured, it could stem bleeding, maintain pressure, even do chest compressions if necessary. In fact, once such a second skin becomes widely adopted, life without it may quickly become unimaginable and scary.

Wearable technology may become the ultimate in small or portable technology because of the convenience and effectiveness of being able to carry it around with us. As shown, many of the technologies we are discussing might converge on wearable technology, which is a good reminder that when we try to imagine the future, we cannot simply extrapolate one technology but must consider how all technology will interact. We may be making our wearables out of 2D materials, powered by AI and robotic technology, with a brain-machine interface that we use for virtual reality. We may also be creating customized wearables with additive manufacturing, using our home 3D printer.

14. Additive Manufacturing

Three-dimensional printing connects the digital and physical worlds.

Much of futurism, including this book, is about the type of stuff we will have in the future. This chapter is about how we will make that stuff. We can start by considering the ideal—what would be a perfect manufacturing process?

Pure creation would be the ability to turn energy directly into matter, of the precisely desired material, structure, and form. Einstein's famous equation, $E = mc^2$, means that energy equals mass times the speed of light squared. Essentially, matter and energy are two sides of the same coin, and matter contains a lot of energy. Therefore, a tremendous amount of energy would be needed to make even a tiny amount of matter. For example, the creation of 1 gram of matter would require the energy in 15,000 barrels of oil.

Of course, this also means matter could be turned into a tremendous amount of energy, so theoretically we could turn matter into energy, then back into matter in the form of the desired object. While this may seem the purest form of manufacturing, the amount of energy wasted would necessarily be huge, and the technical hurdles massive, so this form of production is not likely in our future, at least not anytime we can envision.

In the meantime, we will have to be content with turning raw material into objects—altering matter itself without turning it into energy. Within this paradigm, the ideal process could turn any raw material into any object without limitation—even turning elements into other elements. That, however, would require some form of advanced alchemy that, again, we cannot currently imagine (short of a particle accelerator).

So, let's back up a bit further. Let's assume we are starting with raw material that is at least close to the final chemical structure of the object—if we want something made of wood, we are starting with wood. The manufacturing process may treat the raw material to make it stronger, harder, fire resistant, more ductile, or have whatever properties we want. Some manufacturing processes may also combine chemically to make the raw material (by mixing other raw materials) and put it into its final shape.

How, then, can we ideally go from usable raw materials to a final object with the properties, form, and function we desire? We would like whatever process we use to be precise, without limitations, fast, and cheap. You may have heard the old saying "fast, good, and cheap: pick two." Well, we want all three. How close are we to this ideal, and what technologies might get us there?

A History of Manufacturing

We will never know what the first tool or object crafted by a human relative was. This is because it is probable that our hominin ancestors used sticks and other wooden objects as tools (given observations of our chimp cousins), and these would be unlikely to fossilize. So, we can only look for the first evidence of toolmaking among hominins.

The oldest stone tools go back 3.3 million years, in the Lomekwi 3 archaeological site in Kenya. They are large, hard stones clearly worked, although how they were used is unclear. This site predates our own *Homo* genus by about 500,000 years, making it possible that our *Australopithecus* ancestors made stuff, but toolmaking really took off with the *Homo* genus, starting with *Homo habilis* (literally, "handy man").

For about 3 million years, then, shaping stone tools was

cutting-edge (literally) technology. The oldest bone tools appear 1.5 million years ago, also in Africa, but this technology rapidly rose in prominence with Homo neanderthalensis and *Homo sapiens* (us), starting about 150,000 years ago. This is the "stone knives and bear skins" era of human technology. Bone was used to make awls, needles, spear points, and fishhooks. The appearance of needles also implies the use of leather and fur to make clothing. So, the earliest manufacturing methods involved subtractive processes (taking stuff away) to minimally modify raw materials for a specific purpose.

Shaping through subtracting was therefore the first and oldest manufacturing technique people employed. The next was treating a material to alter its properties—namely using fire to harden wood, which started about 120,000 years ago. The next leap in manufacturing techniques involved joining two or more pieces together. The oldest evidence of this is 72,000 years ago, when spear points were joined to shafts of wood using resin and other material.

Another process added to the manufacturing repertoire was coating the surface of an object with some kind of material to change its surface properties. The evidence for this includes cave paintings starting about 30,000 years ago.

A huge leap in manufacturing came about 25,000 years ago with the use of pottery made from bone dust and clay. This is the oldest evidence of shaping or forming material. The clay was also heated to harden it, and so this was perhaps the most significant of the manufacturing processes that changed the properties of material.

These basic techniques—subtractive shaping, joining, treating, coating, and forming—would remain the mainstay of manufacturing until modern times, continuing to this day, and will likely continue far into the future. The choice of materials has increased (as we saw in chapter 11 on materials) and the techniques have evolved, but the basic concepts remain the same.

While materials and techniques advanced, the manufacturing process remained largely the same—individual crafters would create items individually and largely by hand. As the number and complexity of objects increased, crafters became increasingly specialized. It might take years, or even decades, to master a specific craft. Eventually we have woodwrights, blacksmiths, weavers, glassblowers, pottery makers, coppersmiths, leatherworkers, and much more. In large cities where labor was plentiful, specialization would lead to division of labor, where multiple crafters would focus on one aspect of a process to create a single item.

The next biggest revolution in manufacturing came with the mechanization of making stuff. This was most significant during the Industrial Revolution but using machines to help in crafting goes back much farther than this. The lathe, for instance, goes back at least to ancient Egypt, 1300 BCE. It is a device that turns a piece of wood around a radial axis, so that the wood can be carved, sanded, treated, or coated. This creates an object with symmetry around that axis. Lathes can be driven manually but were also connected to wind turbines or water wheels. Other forms of early mechanization in manufacture include the pottery wheel, the sawmill, and the trip hammer used in blacksmithing. Meanwhile, spinning wheels and looms revolutionized the textile industry.

Of course, the Industrial Revolution served as a big leap forward in mechanization of manufacturing, beginning with the invention of the steam engine in 1769. This not only powered existing manufacturing tools, but it also opened up new possibilities in terms of replacing human power with machine power and automating manufacturing processes. With machines it was now possible to mass produce items, to utilize an assembly line, and to have precise interchangeable parts. The trade-off was often that each individual item was not of the same quality as a handmade version crafted by a master.

The cost for machine-produced products could often be orders of magnitude lower. As manufacturing technology advanced, the baseline quality of mass-produced objects also improved. We are now at the stage where items come in a continuum of qualities—for many things, you can purchase a cheap mass-produced version, or increasingly high-quality and expensive versions, to meet your resources and needs.

This current state calls to mind an encounter I had with a vendor at a Renn faire, which often offers goods produced using traditional craftmanship. I was eyeing a wool sweater, and the vendor explained to me that she raised the sheep herself, harvested the wool, dyed the wool, spun the fibers into yarn, and then hand weaved the sweater from that wool. She used these details to justify the rather shocking cost of $3,000. Upon hearing the price, I couldn't help thinking to myself, *That's why we don't do it that way anymore.* Sometimes having an item crafted by a master is worth it. Sometimes that's just doing it the hard way.

Since mechanizing and automating manufacturing in the nineteenth century, have there been any other significant manufacturing advances? Absolutely. One aspect of manufacturing progress is in design—how we go from a concept to a final product. The oldest technique for design was simply the mind of the crafter. Hand and skill were used to convert a mental image into a physical object. This technique could also be supplemented by copying an existing object, or copying elements from nature, by direct visual inspection.

As crafted objects became more complex, more precise techniques were necessary to create them. This is where patterns and blueprints come into play. You can record the precise details of an object's construction for a crafter to follow, but this might also include a physical reference that actually controls the crafting process—a template.

Casting or mold making is another way to control the shape of an object. The oldest copper cast item known is a frog dating from 3200 BCE from Mesopotamia, although evidence for casting goes back to about 4000 BCE. Injection molding is the process of injecting liquid plastic material into a mold that can be repeated to mass produce items of the desired shape. The first patent for an injection molding machine was issued in 1872 to John Wesley Hyatt (remember the guy who helped invent the first plastic?). As of 2018, plastic injection molding is a $139 billion annual business worldwide, and still growing by about 10 percent per year.

The huge paradigm shift in design of controlled manufacturing was the shift from physical templates and molds to programmed information, which of course required computer technology. Computers now allow for the digital engineering and design of objects and increasingly direct the manufacture of those objects.

There are two basic ways to get a digital "blueprint" or pattern for an object. One is computer-aided design (CAD) where computer software is used to create the design. CAD software includes precise drawing and drafting tools to create all the information necessary to generate the desired object. The other main method is scanning—starting with a physical object and then using high-definition scanning technology to create a digital representation. These methods can also be combined.

There are a number of ways that this digital design can connect to physical manufacturing. Robots can be programmed to carry out manufacturing procedures to create the final product. The current cutting-edge technology for this purpose is computerized numerical control (CNC) machines, which are controlled by computer software to operate lathes, drills, laser cutters, routers, grinders, or any other machining technology in order to produce digitally designed items.

CNC machines operate in either an open loop or closed loop. In open-loop systems, the software directs the machines according to a predetermined design without any feedback. In closed-loop systems, the CNC software can receive feedback, such as images, from the item being created, to correct errors or rectify any irregularities. Of course, artificial intelligence is now being incorporated into these closed-loop systems, making them more adaptive, efficient, and precise.

CNC technology is already common in most manufacturing industries and growing rapidly. It is likely to remain a critical component of manufacturing for decades. However, the CNC process is a form of subtractive manufacturing—you start with a block of raw material that is removed in order to make the final product—and there is an inherent amount of waste. Future competition will come from additive manufacturing, building up to the final product rather than taking away.

The latest version of additive manufacturing is 3D printing, which operates similarly to a regular printer but in three dimensions. This technology represents a direct connection between the digital design of an object and its creation—you are literally just printing the design. Yet again, this technology has deeper roots than you might imagine.

The basic concept of creating a 3D image of something to use as a template goes back to the 1800s. In 1859, François Willème developed a method called "photographic sculpture" in which he could capture 3D models of human subjects using twenty-four cameras placed at different angles. Joseph E. Blanther patented an apparatus that used layering to create three-dimensional topographical maps in 1892.

As for 3D output, in 1980, Hideo Kodama of Nagoya Municipal Industrial Research Institute invented a single-beam laser curing approach for a rapid prototyping system. He applied for a patent in Japan in May 1980 but did not have the funding to see it through. His system would use ultraviolet light to harden parts of a photopolymer into the desired shape.

Between 1984 and 1986, French scientists Jean-Claude André and Alain le Méhauté developed a system for stereolithography. This used two laser beams controlled by a computer design, which converted a liquid monomer into a hard polymer where they crossed. They built this system to demonstrate principles of fractal geometry.

At the same time, engineer Chuck Hull also developed a 3D printing system using UV light and stereolithography. His purpose was also rapid prototyping, but he saw that such a technology would also be popular in the growing maker culture (a do-it-yourself subculture for making all kinds of things).

Then in 1988, engineer Carl Deckard developed what he called SLS—selective laser sintering. Sintering is the process of heating small pieces of metal until they become soft and sticky in order to combine them together. Deckard was working in a machine shop and thought that doing so many metal castings was too cumbersome, and therefore wanted to develop a method for rapidly producing metal parts.

In that same year, Scott Crump developed what he called "fused deposition modeling." He connected a glue gun to an XYZ 3D gantry in order to automate the modeling process. His gantry system is still the basis of about half of modern 3D printers. Clearly this was an idea that was ripe—the connection of the burgeoning computer technology with manufacturing.

In the late 1980s, 3D printers were used mainly for prototyping in the automotive and aerospace industries, with Ford and Boeing being early adopters. They allowed for much more rapid turnaround time in terms of design. Rather than waiting months to get and test a prototype, you could have it in hours.

In the next decade, advances to this technology spread to the medical industry. Bioprinting by layering cells on top of each other started, and in 1999 researchers at the Wake Forest Institute for Regenerative Medicine used bioprinting to make an artificial bladder using the patient's own cells.

In 2004, the first desktop 3D printer, the RepRap, was created, with a commercial version being released in 2007. From that point forward, like most technologies, 3D printing continued to improve, with a greater range of materials, faster print times, higher resolution, and larger build areas. The home desktop 3D printer market, however, crashed in the 2010s. The reality did not live up to the premature hype for many early adopters.

Today the 3D printer market is used for rapid prototyping, for a small market of makers, and increasingly for actual production in industry. At this point, "3D printing" is often used as a synonym for all additive manufacturing. There are several methods for 3D printing, involving turning some liquid or soft material into a hard material that is layered up using a print head that can move in three dimensions.

The 3D printing industry was worth about $13 billion in 2020, with annual growth rates of about 26 percent. The home market remains small, with inklings of growth as printers become less expensive and their capabilities mature.

Three-dimensional printing is not without some limitations. While it uses a process of addition, that does not mean there is no waste. Some shapes are inherently unstable (they don't stand on their own) and so supports have to be constructed and then removed. There are already methods to minimize this, through good design planning and printers with movable platforms that reduce the size of the needed supports. The surface characteristics of printed objects may also not be as clean as CNC production and therefore need postprinting processing.

If 3D printers can create something without waste or flaws, they may not be in the preferred material, as not all are yet available—for example, metals with high melting points. Three-dimensional printing is also mostly a monochromic process. You can't 3D print in full color.

Given these are technological limitations, though, there is no reason to think they will not be overcome with continued incremental advances in the technology.

How Will We Make Stuff in the Future?

The future of manufacturing is clearly digital. Older technologies typically do not completely disappear. They persist with amazing resilience, but when they are displaced by superior technology, their niche becomes progressively smaller. Traditional handcrafting of objects will long have a place for their perceived quality, uniqueness, and attention to detail.

Mass production using physical methods like molds and templates will likely also persist for the kinds of objects and materials where those methods are best. If you want to make millions of identical small plastic toys, for example, injection molding is still the top option.

Since the ability to connect a digital image or design of an object to a faithful production of that object is extremely powerful, the future is therefore likely to be built mostly with some combination of computer-controlled robotics, CNC machines, and additive manufacturing.

There are already factories with rows of industrial 3D printers for rapid design and production of items, and this will likely increase. Imagine a massive factory with a million industrial-sized 3D printers, able to mass produce almost any item on demand. There is no need to retool the factory, so turnaround time is instantaneous.

Three-dimensional printers also already have a foothold in remote locations and niche industries. Whenever you need to rapidly create a unique item, a 3D printer is ideal. In the medical industry, 3D printers can be used to make braces, implantable devices, and prosthetics based on MRI scans or other imaging. Surgeons can also 3D print models of their patients' anatomy to practice or design the necessary surgical procedure.

One big question remains, however—how much will 3D printers come into the home? There will always be technophiles, or hobbyists, who will want specialized equipment. But will the 3D printer become a ubiquitous home appliance, like a refrigerator or microwave? When trying to predict such future questions, it is important to think about what the technology will be competing with. In most cases, the technology itself is not the question—3D printers exist, they are within average consumer pricing, and they are only getting better, but will they be worth it for routine use?

The competition in this case in centralized production. Why maintain and operate a 3D printer when you can order anything you want online and have it the next day? Centralized production benefits from an economy of scale, and higher end industrial printers. It's also quicker for the user, and time is always a premium commodity. It does seem, however, that at some point 3D printing technology will become so easy and cost-effective that it will outweigh the barrier to owning and operating one.

With a 3D printer in every home, you will only need raw material and digital designs. There are already millions of free digital designs available as open source, and more being added every day. By the time 3D printing at home becomes common, you will likely be able to download the design for any imaginable common object. Specific pieces, such as a replacement for a broken part, could be obtained from the manufacturer. In fact, they would probably come with the item—if this doodad breaks, just print a replacement using this design.

The real intellectual property will be in designs themselves, for art, upgrades, or unique items. In this world, everything could be personalized.

Perhaps even more important than improvements in 3D printing hardware is improvements in the CAD and printer control software, which will get a huge boost from AI technology. Imagine interacting with your 3D printer in natural language, and it making almost anything you want. You don't have to be an artist, an engineer, or an expert at using complex software. The AI will sort out all the details.

Centralized manufacturing will then shift to distributed manufacturing, at the point of use. Likely there will remain complex and specialized items (like circuit boards) that will have to be produced in a factory, but any simple or solid object will be printed at home.

For the farther future, there are at least two technologies that could compete with and even displace digitally controlled additive manufacturing. One is the aforementioned programmable matter, which can reconfigure itself to conform to the digital design. Beyond that is mature nanotechnology—rearranging matter at the molecular and even atomic level, to create almost anything from anything. With mature nanotech, we are getting close to our ideal manufacturing, if we can take the steps to get there.

15. Powering Our Future

We are crafting the energy production of tomorrow with the decisions we make today.

As we try to imagine our technological future based on past and present technology, we have taken a deliberately broad approach. In that vein, for the last chapter of this section, I would like to consider the broadest perspective I can. What are the primary, most important, or most limiting resources that human civilization depends on?

Material resources are certainly significant. We need to build our civilization out of something, and we always seem to be facing a crisis of "peak" whatever—peak oil or peak helium—as our demand for a resource begins to outstrip our supplies.

Land is also in limited supply. There is only so much land on Earth on which to grow our food, build our homes, and contain our industries. It would also be nice to leave some land for the other 10 million or so species that we share the planet with.

I would argue, however, that the primary resource, the one resource to rule them all, is ideas. With science and technology, we have, so far, been able to overcome all our other resource limitations. We need to grow more food, so between 1948 and 2017, we nearly tripled the productivity of farmland through advances in agricultural technology. Eventually we'll be harvesting material from asteroids, expanding onto other worlds and deep space, and maybe even reversing some of the damage we have done to the natural world.

This view supports the notion that perhaps AI is the most important technological advancement—increasing our most vital resource, thinking. In that case I would argue that the second most valuable currency of our civilization (in fact, for the universe) is energy. As long as we have a source of energy, we can harness our ideas and accomplish almost anything.

In fact, in 1964 Soviet astronomer Nikolai Kardashev proposed a classification system for measuring the level of technological advancement of a civilization based entirely in how much energy they are capable of using. A type I Kardashev civilization can harness the energy of its entire host planet and is therefore considered a planetary civilization. Type II is a stellar civilization, harnessing the total energy output of an entire stellar system. And type III is a galactic civilization, controlling the entire energy of a galaxy.

On this scale, human civilization would not even rank type I, and for this reason the scale has been extended. Carl Sagan proposed a system for mathematically extending the scale, making a type 0 civilization one that controls 1 MW (megawatt) of energy (enough to power about 420 average American homes). On this scale human civilization would rank about a 0.7. Physicist Michio Kaku later argued that if we continue to increase our energy use by about 3 percent per year, we will achieve a type I civilization in 100 to 200 years. I wonder if we'll get a badge or certificate or something.

If we hope to someday gain entry to the type I civilization club, how will we get there? We are currently engaged in a heated debate about the future of our energy infrastructure. We must meet growing energy needs in a sustainable way while having a lighter footprint on the environment. There are multiple paths we may take to get there, but first let's consider the basic concept of energy itself. Where does it come from, and how can we harness it?

If you trace it back, the sun is the ultimate source of most (not all) of the energy we use. The sun bathes Earth in radiant energy, and the biosphere has been soaking up that energy for billions of years. Plants have learned how to convert that solar energy into biochemical energy through photosynthesis. Other creatures eat those plants, and are in turn eaten by other animals, and when they die, their decomposition feeds other organisms. The (almost) entire biological ecosystem is therefore driven by solar energy.

The tiny exception to this is chemosynthetic organisms that feed on geologically produced methane or sulfides that leak out of ocean fissures. Chemosynthetic organisms may, in fact, be dominant on other worlds where sunlight is minimal or absent, such as below the ice of Europa.

So, in a way, humans' first energy source was biochemical, fueled ultimately by sunlight. The technical definition of energy is the capacity to do work, and so the work of human civilization was largely from muscles burning calories that came from food whose energy came from sunlight.

The Wonderwerk cave in South Africa's Northern Cape province holds the oldest direct evidence of fire use by humans, going back 1 million years. However, there is indication that cooking with fire goes back to *Homo erectus* about 2 million years ago. While the earliest use remains debated, it is clear that the use of fire by humans was widespread as far back as 400,000 years ago.

Harnessing fire was transformational for our human ancestors. Fire provides light and heat and can be used to fend off predators. Perhaps most importantly, it allowed for the cooking of food, which not only expanded our food choices but also made it much easier to digest food and access its nutrients. Scientists credit cooking with the huge expansion in brain power (the brain is a very hungry organ) from *Homo erectus* to *Homo sapiens*. It literally made us who we are.

Even with all our modern technology, fire remains an important source of energy for human civilization. At first, we depended entirely on combustible material from living things, like wood, natural oils, and wax. We did find some petrochemicals (literally "from rock" like bitumen), but it wasn't until 1875 that crude oil was discovered by David Beaty at his home in Warren, Pennsylvania, changing our energy infrastructure for the next century and a half at least.

Coal, which also goes back thousands of years, has the benefit of being more energy dense than wood, so you get more energy per volume. On the other hand, it has a serious disadvantage of releasing pollutants in the air (some worse than others, depending on the type of coal). Collectively coal, oil, and gas are referred to as "fossil fuels." This is because they originate from biological material crushed over millions of years by geological forces after dead things get buried in the ground. Therefore, fossil fuels are stored solar energy.

It is also worth noting that burning itself is simply a version of chemical energy. It is the combining of a combustible material with oxygen in a self-sustaining exothermic reaction. Some of the chemical energy for that reaction is contained in the free oxygen itself, which is a highly reactive element. Where does the free oxygen on Earth come from? It also comes from life. Cyanobacteria and other photosynthetic organisms use sunlight to strip oxygen from water and carbon dioxide, releasing free oxygen into the atmosphere. When we burn fuel, we are combining one set of compounds that derive from sunlight with another element, free oxygen, produced using sunlight. It's all solar energy.

What about wind and water energy? Even before the harnessing of electricity, wind power was used to turn grindstones to mill wheat and other grains (windmills). Wind power could also turn gears to do other work, like turning a lathe for woodworking or supplying air for a forge.

However, wind results from differential heating of air—by the sun. A little is kinetic energy from the rotation of Earth, but this is a minor component. Therefore, most of wind energy is also solar energy.

Water power could do the same things as wind, by building a water wheel on a river with a natural gradient. You might think that water power is ultimately driven by Earth's gravity, which is the force that makes water fall. This is true—as far as it goes. The potential energy of the water is converted to kinetic energy as it moves down the gravitational well, and some of that kinetic energy turns the turbine. But where did the water get its potential energy from in the first place? Water evaporates when heated by the sun, collects in clouds, then rains down where it collects in lakes and then flows down to lower altitudes. So hydropower is also just another form of stored solar energy.

Are there any sources of energy that do not derive ultimately from the sun? A few. Geothermal energy comes from the internal heat of Earth, which, like all the planets, formed through the collision of smaller rocks. This bombardment converted a tremendous amount of gravitational energy into heat. About half of the geothermal energy near Earth's surface comes from the heat of radioactive decay from radioactive materials in the crust—uranium, thorium, potassium, and others. So geothermal energy is partly gravitational and partly nuclear.

And, of course, nuclear power also comes from radioactive materials. Nuclear decay can be directly turned into electricity or used for heat. Fission (breaking apart) of radioactive isotopes can also be used to generate large amounts of heat to generate electricity.

Another potential source of energy is tidal—the gravitational fields of the moon and sun cause the oceans to rise and fall, by as much as forty feet in some locations. Europeans harnessed this movement to power mills a thousand years ago. Today this movement can be harnessed to generate electricity.

Finally, there is nonbiological chemical energy. While batteries, for example, can be used to store energy, they can also be a (very limited) source of energy if made from chemicals that will react to generate ions that move from the cathode to the anode. Chemical batteries can therefore be a way of harnessing chemical energy. For nonrechargeable batteries, this reaction moves in one direction. In rechargeable batteries, the reaction can be reversed by the application of an electrical current.

These, then, are the major sources of energy that we use to run our civilization. Our relationship with energy, however, dramatically changed with the use of electricity. Electricity comes from harnessing the electromagnetic force to create a current of electrons in a circuit. That flow of electrons can be made to do work, such as lighting a light bulb or running a motor.

While many people think of Benjamin Franklin as the person who discovered electricity, it was probably the ancient Greeks around 600 BCE—they found that they could cause amber and wool to be attracted to each other if they rubbed them together (static electricity). There are also ancient Roman and Persian artifacts that look like primitive batteries, made from iron and copper rods in a clay jar. Our understanding of electromagnetism really advanced in the seventeenth century, ultimately leading to the electronic revolution. By the early twentieth century, the electrification of the modern world was well underway, and now it seems inseparable from modern technology.

Electricity is mostly generated by turbines. This is based on the physical principles of electromagnetism—an electrical current will generate a magnetic field, and a moving magnetic field will generate a current in a conductive material. So, you can spin magnets around a wire to generate current, or you can use an electrical current to turn magnets and generate motion (which is a motor).

A variety of energy sources can be used to turn a turbine such as wind (wind turbine) or moving water (hydroelectric power or tidal power). Heat can also be used to boil water, the rising steam from which can turn a turbine. That heat can come from geothermal energy, solar thermal energy, burning wood, coal, natural gas, or oil, or from nuclear fission. In the end they all work the same way.

Solar power from photovoltaics is different. It generates an electrical current directly through the photovoltaic effect—a photon of light hits a material that knocks off an electron, which generates current.

Even radioactive decay can be used to generate electricity directly, through radiovoltaic conversion, or indirectly through heat, light, or electrostatic production. These currently have limited uses, such as in satellites and probes, but have been considered for more general use (remember those nuclear-powered vacuum cleaners that were proposed in the '50s?).

Power Production Today

Figures from 2021 published in the *Economist* indicate that the world mix of energy sources was coal 29.6 percent, natural gas 22.5 percent, hydroelectric 2.6 percent, nuclear 5.9 percent, petroleum products 28.6 percent, and nonhydro renewables (mostly wind, solar, geothermal, and biomass) 10.8 percent. That's 80.7 percent total from fossil fuels. The mix can vary considerably for different countries. For the United States, the mix was coal 22 percent, natural gas 36 percent, all renewables 21 percent, and nuclear 20 percent (for a total fossil fuels of 58 percent).

In the United States, renewables, especially wind and solar, are rapidly trending up. Natural gas has recently increased significantly, but it is now slowly trending down. Coal is rapidly trending down. Nuclear is holding its own due to extending the license on existing nuclear power plants.

The question becomes what is our energy infrastructure going to look like by 2050, 2100, and in the far future? As we have seen with future technology in general, it all depends on the choices we make today. There are numerous factors to consider, including a couple more aspects of energy infrastructure we have yet to discuss: distribution and storage.

In addition to the source of energy, we need to consider how it will be distributed. Right now, electricity is generally centrally produced and then broadly distributed along large grids (essentially a managed web of transmission lines). In the lower forty-eight states of the United States, for example, there are three grids—an eastern grid, a western grid, and Texas. Texas remains on its own grid to avoid interstate regulations. The largest grid in the world is the synchronous grid of Continental Europe, which serves twenty-four countries.

The advantage of large grids is that they allow for sharing energy, which makes it easier to average out peaks in demand or compensate for a loss of production. Larger grids will also make integrating intermittent sources of power, like wind and solar, easier. The wind is always blowing somewhere, so a widely connected grid could essentially turn a sporadic source of power into a reliable one. Sharing solar power would require very large grids that can transport power from sunlit parts of the world to dark parts of the world, something there are no plans for at the time of this writing.

There are a lot of calls for upgrading the various energy grids. A modernized grid can be smarter, incorporating computer technology and AI control systems to optimally balance supply with demand and prevent blackouts. Upgrades could also harden the grid against disruptions from coronal mass ejections or even attack.

Further, an upgraded grid would better allow for distributed energy production. This is one trend in energy infrastructure that can have a huge impact on the future. Instead of generating power centrally and distributing it, a larger portion of energy can be produced locally. This has several advantages. First, the power does not have to travel as far, so there is less loss in transmission. Currently an average of about 6 percent of generated electricity is lost to transmission, and staying at this number requires the use of high voltage to minimize loss, which creates safety issues. Local production would also decrease the probability of power disruption from downed lines or other problems.

Energy production can also produce a lot of waste heat, requiring cooling, but locally produced energy can use the waste heat for heating or even for manufacturing. This could significantly improve the overall efficiency of energy production, especially with turbine-based electricity production where about a third of the energy is wasted as heat.

Local energy production, however, is limited by the grid, which needs local transformers (which change the voltage of the electricity) to accept the energy. Current grids are simply not set up for massive distributed production. The number of houses on one street, for example, that can have rooftop solar attached to the grid is limited, and it can be costly (and fall on the consumer) to add transformers that accept more rooftop solar.

The addition of grid-level energy storage to our current system is therefore an important factor for our future energy infrastructure. There are many types of grid storage: batteries, pumped hydro, heating salt rocks, spinning up flywheels, hydrogen, and compressed air. Of these, pumped hydro and lithium-ion batteries are the most energy efficient (referred to as "round-trip" efficiency—how much energy is lost converting electricity to its stored form and then back again to usable electricity). There is also research into new methods, such as stacking giant concrete bricks to store the potential energy.

Grid storage is being added at a rapid pace around the world, but it still accounts for only a tiny fraction of energy demand. Storage allows for rapidly dispatchable energy—energy that is available immediately to balance supply and demand—and for more use of intermittent sources, such as storing energy while the sun is shining and then using it during peak demand after sunset.

Storage is a huge advantage for utility companies that have to keep enough energy capacity on hand to manage peak demand, even if they only need to produce it for an hour or so a day. Turning power plants on and off is inefficient, and companies further reserve their least-efficient energy sources for a last resort when demand peaks. Grid storage can be used for "peak shaving"—meeting those peak demands without having to fire up that natural gas power plant.

Where Are We Headed?

Assuming no disruptive technologies are developed (we'll save that for later), and we continue to have the same rate of incremental improvements in existing technologies to plausible theoretical limits, what will our energy infrastructure likely be by 2050? One path we might follow, the one we are largely following now, is to simply let market forces determine our energy mix. If this is the case, then we will likely continue to pull fossil fuels out of the ground and burn them. This is especially true in the developing world, like China, which is currently burning more coal than any other country.

But fossil fuels are no longer the cheapest source of new energy. They are getting more expensive as reserves and sources dwindle. Wind and solar are cheapest, and getting significantly cheaper, which is why they are the fastest growing segments of the energy sector. Solar especially is not only the cheapest current form of electricity, but it is also the cheapest energy of all time. If you are a utility company and you need to add energy capacity, building wind and solar is your cheapest bet. Hydroelectric and geothermal are also cost-effective, but they are geographically limited, which restricts their potential to maybe 10 percent of our energy needs.

The technology of wind power is also improving, making it more efficient and quieter, while also reducing its use of limited resources like rare earth elements. There is, for example, a vertical wind turbine design where instead of a pinwheel arrangement, the blades are all vertical and rotate around the central support. They are more efficient and can be arranged together to further enhance effectiveness.

Nuclear, however, is an uncertainty. It suffers from having the most expensive start-up costs, but all prior predictions of nuclear energy's demise have been premature. The industry is responding with designs for small modular reactors that are much less expensive to build. Further, there are plans for "Generation IV" nuclear power plants. These are much safer than older designs, some are incapable of melting down, and they produce less waste. In fact, several Gen IV designs are fast breeder reactors capable of burning reprocessed spent fuel from older reactor designs.

Current nuclear plants use only 5 to 6 percent of the energy in the uranium fuel. The rest becomes waste with half-lives in the thousands of years that need to be stored. Fast reactors can burn that waste, extracting 95 percent of the energy from the uranium fuel, and what is left has a half-life in only the tens of years. There is enough current nuclear waste to supply the energy needs of the United States for the next century using these Gen IV designs, without having to mine any further uranium.

If market forces are the only indication of the future, then I suspect we will continue to see nuclear power, mostly in developed countries, and running between zero and 20 percent of production (depending on public acceptance).

In addition to that, how far will wind and solar go? The greater penetration they have into the energy sector, then the greater the need for an upgraded grid and for grid-level storage. So, this will depend on the willingness of countries to pay for infrastructure. These intermittent sources also become less cost-effective, as their penetration increases because they require overcapacity. You have to build more wind turbines to make sure some are functional at all times, which means that the percentage of time any given wind turbine is generating electricity will be lower. Also, once premium locations are used up, they will move into less-ideal locations.

Market forces will therefore likely favor a continuation of nuclear at levels similar to today or a slow decline, wind and solar somewhere around the 50 to 60 percent mark, other renewables like hydro and geothermal between 5 and 10 percent, and the rest natural gas mainly used for dispatchable power. Developing countries will likely continue to use coal. The required significant grid storage will likely be mostly batteries. This represents a fairly straight extrapolation of current trends. There is some question about how fast we will get there, but thirty to fifty years is reasonable.

However, market forces are not the only factor at work. The big elephant in the room is concern over climate change. Countries are increasingly considering the release of CO_2 to be an externalized cost of fossil fuel. Economic studies predict the cost of climate change may run in the trillions of dollars by 2050, and we could be seeing a climate refugee crisis.

Fossil fuels also cause health-harming pollution. The health care costs alone (which is also an externalized cost) are billions of dollars a year worldwide, and something like $100 million per year for just the United States. Specifically, burning coal releases more radioactive material into the environment than nuclear power and has the highest death count per energy produced.

What if, instead of getting subsidies, which were estimated to be about $1 trillion worldwide in 2021, fossil fuel companies were made to pay the externalized cost of burning their products? These political choices can drastically affect the energy market, making fossil fuels much less cost-effective and favoring renewables and nuclear power.

The fate of nuclear is also very much a regulatory question. Much of the cost and delay of building new nuclear is because there is still widespread public opposition to nuclear power lingering from the antinuclear movement of previous decades and legitimate concerns about storing nuclear waste. The scientific community remains more supportive of nuclear energy, especially the Gen IV designs. Nuclear waste is entirely manageable, and in fact can fuel the next generation reactors. Nuclear power is also one of the safest forms of energy, far safer than any fossil fuel, and is comparable to renewables. It also has the least carbon footprint of any energy source.

So, energy production in 2050 to 2060 could look very different depending on the choices we collectively make, but I suspect it will mirror my extrapolation above. Our choices mostly affect how quickly we phase out fossil fuels, not if we phase them out. They will also determine how long we keep nuclear fission in the mix. Interestingly, therefore, our energy infrastructure over the twenty-first century may have more to do with our political decisions than technological advances, which have been relatively steady and predictable.

What about those disruptive technologies, or radical approaches to producing energy? Say we wanted to maximize our solar energy production; we could not only maximize distributed rooftop solar but also create massive solar farms in desert environments.

It has been estimated that if we cover 1.2 percent of the Sahara Desert in solar panels, representing 112,000 square kilometers, or an area of 335 × 335 km, we could power the world entirely with solar power. This, however, is impractical, as it would also require an incredible worldwide grid with enough storage to power the world overnight. That's why this is more of a thought experiment to estimate how much solar it would take to power the world.

We could, however, more feasibly put hundreds of solar farms in deserts around the world, although it still requires massive grids with lots of grid storage. While deserts may seem to contain little life, though, they have ecosystems of their own, and very large solar farms can disrupt them. This is why many experts recommend that we don't try to meet our energy needs with any one solution, but by combining all viable low-carbon solutions. We can pick the low-hanging fruit from each energy source, without trying to have a one-size-fits-all solution.

Another radical idea involving solar, however, is orbiting solar panels that collect energy and then beam the energy to stations on Earth. This is more plausible than it may sound. Being in high orbit means that solar panels can get sunlight twenty-four hours a day, roughly doubling their efficiency, and there will never be any clouds in the way. We would need something like 70,000 square km of such panels—still a lot, but there is more than enough room in geosynchronous orbit. Energy could be beamed down to receiving stations around Earth as microwaves or lasers.

The downside to such a system is that getting objects into high orbit is extremely expensive, currently about $1,500 per kilogram at the low end. As we will see in the chapters on space travel, there are efforts underway to bring this cost down significantly, but those plans would have to succeed before space-based solar becomes economically feasible. Once in orbit, geostationary satellites are also difficult to get to, and therefore to repair. It would be necessary to consider the life span of each satellite and the cost of replacement in any cost analysis. This may be an energy solution for the next century, however.

Despite solar roadways gaining a lot of attention online, I am fairly confident that we will not be seeing them as a serious energy solution anytime—ever. It simply makes no sense to put solar panels on roads. They would have to be extremely durable and resist scratching and scuffing. They would also need to be plowable. There's simply no reason to drive on your solar panels.

What other options are there beyond a massive investment in solar energy? It may be noticeable that I have not mentioned hydrogen power so far in this chapter, except a nod to grid storage. That's because hydrogen on Earth is not a source of energy. Free hydrogen is, but there is essentially no free hydrogen on Earth. In the future, if we can mine the atmosphere of Jupiter for free hydrogen, things may be different. Hydrogen may become the new "fossil fuel" that we suck up and burn.

As long as we are trapped on Earth, however, we have to split hydrogen from water or another hydrogen-containing molecule. This uses up energy, some (but not all) of which is recovered when the hydrogen is burned with oxygen to make water again. So, hydrogen can be used to store energy and generate energy using hydrogen fuel cells, but as we discussed in chapter 10 on transportation, hydrogen fuel is not as efficient as lithium-ion batteries. For this reason, it seems that batteries are winning the EV war, but hydrogen will likely have a small niche market where its fuel cells work best.

There is a lot of research into "artificial leaf" technology, which essentially uses sunlight to split hydrogen from water. If this can be done efficiently, then hydrogen may become a major way that we store solar energy.

Another possible new entry into the energy mix is thorium energy, which is a form of fission that many consider to be safer (and less weaponizable) than uranium. There is a working thorium reactor in India, but overall, this technology lost out to uranium cycles. This may change at some point in the future, especially if uranium supplies dwindle. Thorium is more abundant in different parts of the world, such as the United States, so its use may be regional.

Biomass or biofuels also deserve a mention. There is a lot of biomass (which just refers to any biological material, like plants) that is essentially waste, and converting it into fuel that can be burned can add to the energy mix. I doubt this will become a significant percentage because we are already maximizing our land use just growing food let alone fuel. However, there is research into growing bacteria or yeast in large vats that produce chemicals that can be turned into fuel. If this process can be made energy and space efficient, it may have a niche for certain applications, like jet fuel, but is unlikely to be a major contributor to our power mix.

While these are all the plausible sources of energy that we are likely to have anytime soon, there is one more extremely disruptive energy technology, fusion, that could be on the cusp. We ultimately deemed it a future technology that does not yet exist and will be covered when we consider what new technologies might come into existence that would change our world.

FUTURE FICTION: 2209 CE

Ling walked the backstreets of San Francisco with the comfortable ease of someone who had lived there her entire life. She knew how to avoid the countless drones, surveillance cameras, and visibots. City life could never be free of such intrusions, but there were places where one could duck out of view.

Grif, her lone companion, flitted about. He was getting big, and Ling wondered if his flying would eventually be compromised. The wings she designed for the 10 kg feline were large, but if he put on even one more kilo, he would no longer be able to leave the ground.

If she won today's purse, then she could afford some upgraded genetic algorithms and more growing medium. Even then, would she splurge on her companion? She still had so many upgrades planned for herself. She was especially looking forward to increasing her neuronal density, but that was a pricey one.

At the right moment, she ducked into an alley, down a short staircase, and through the open doorway. Her AR contacts were the only wearable technology she allowed herself, and even then, she wore them with a twinge of purist-biohacker guilt. They proved their need as she saw the virtual sigils on the grimy walls. They let her know she was in a surveillance-free zone, that she was headed in the right direction, and the codes she would need to get past the AI guardians along the way. Besides, she couldn't compete without them.

Finally, she emerged into the staging area, with about fifty people turning to her simultaneously. Was she late? The AR clock told her she was exactly on time, but everyone looked as if they'd been here for ages. The crowd was easy to sort, with biohackers on one side and cyborgs on the other. She joined her group and was met with encouraging pats from clawed hands, relieved looks from vertically slit eyes, gentle hugs from prehensile tails and not-so-gentle hugs from the more muscle-bound hackers.

Grif jumped up on his usual perch, and Ling made her way to the center to face her opponent, the other finalist. The two looked each other up and down, almost making a show of assessing what they were up against.

Her opponent's racing name, Sleek, was apt. She was about two meters tall and slender, dark-skinned with no visible hair. At first Ling thought she was wearing a helmet, but on closer inspection it was an interface with likely AI augments. Her limbs were more composites than flesh, and she had something on her back that was likely an auxiliary power source.

Sleek looked over Ling with glowing cyborg eyes. "You don't look like much," she said, engaging in the smack talk that was as much a part of the tradition as the event itself. "I was expecting at least a tail." She shared a derisive laugh with her cyborg fans.

Ling ignored her, deciding to play it cool, and gave only a dismissive shrug.

An older, apparently unmodded gentleman stepped into the middle holding a device that Ling was all too familiar with. He said nothing, just gestured to the racers. They each held out their forearm and the man placed the device against them in turn, leaving behind a softly glowing tattoo.

"You're now tracked," he said in an incongruently high voice. "You know how this works—you have to hit all three waypoints. First one back here wins. Other than that, anything goes." He emphasized the last words, and the crowd predictably responded with hoots and cheers.

Ling's AR added a red filter to her vision as details for the day's race appeared before her—the location of the waystations, the purse, the stats of each contestant, best times by prior competitors.

There were two doorways leading away, one marked with an image of her face.

Suddenly the red filter turned to green, and the race was on. Both runners bolted through their doors, which led to the same tunnel, stretching at least a quarter mile ahead. Ling pumped her enhanced muscles, with more efficiently arranged myofibrils and densely packed optimized mitochondria. Her legs propelled her with super-strong, elastic tendons. Still, Sleek took an early lead. Ling heard the whirring of her cyborg legs fade slowly as her competitor pulled ahead.

But Ling wasn't worried. These competitions were finely calibrated, and the cyborgs always took an early lead. Ling's real advantage was endurance. The cyborgs always ran low on power first. She just had to stay close enough that she could overtake the lead in the final stretch.

The end of the tunnel rushed up to her, as she scanned quickly for the desired exit. There was nothing but a small square window on the far wall. Ling would have to jump through headfirst without knowing what was on the other side. Sleek was already through, and she didn't hear any screams, so it was probably okay.

At the right instant she leaped forward, diver style with arms out in front, and slipped through the opening...

PART THREE

Future Technologies That Don't Exist (Yet)

One way to explore possible futures is to carefully study the arc of technological history and then try to extrapolate a bit into the future. But this method cannot account for entirely new or disruptive technologies, or changes to society that alter our relationship with technology. For that we need mostly raw imagination, constrained and informed by a basic knowledge of physics, biology, and the other sciences.

New future technologies are harder to predict because they do not yet have a proof of concept. They may never materialize, for various reasons, but if they do, they are likely to be world-changing and have a dramatic effect on future society. We therefore must take a chance and try to imagine which such future technologies are likely to come to fruition.

Many of them started as ideas born of science fiction, and therefore science fiction can be as much of a guide to our future as science. Some of these fantastic technologies will remain in the realm of fiction and may never leave, but either way, we'll discuss how plausible they are, and what impact they will have if they are ever realized.

16. Fusion

Our green future or expensive hype?

Fusion engines are a staple of science fiction for good reason—we will need some way to reliably generate massive amounts of energy on demand if we want to fly around in space, settle new worlds, and run a world awash in digital technology. Right now, we mainly run our civilization on chemical energy and always have, although as I pointed out, much of this energy originated from the sun. The source of chemical energy, from forming and breaking chemical bonds, is electromagnetic. Chemical reactions involve swapping around electrons, and such bonds typically store about 1 eV (electron volt) of energy.

The exception to the dominance of chemical energy is nuclear power, which derives from the strong nuclear force that bonds protons and neutrons in the nucleus of an atom. Nuclear reactions involve changing the number of protons or neutrons in the nucleus, so they can release a tremendously greater level of energy than chemical reactions. These nuclear bonds are in the range of 1 MeV (megaelectron volt), which is 1 million times greater than a chemical bond.

There are, in turn, two types of nuclear reactions, fission and fusion. Fission involves breaking apart large radioactive nuclei into smaller isotopes (different number of neutrons) or elements (different number of protons). Fusion, on the other hand, involves combining small nuclei into heavier elements, such as fusing two hydrogens with one proton each into one helium with two protons. To give you an idea how powerful fusion energy can be, it is the process that powers the sun.

Nuclear fuel is therefore much more energy dense than chemical fuel. A pound of uranium, for example, can produce 16,000 times as much electricity as a pound of coal. That figure could be much higher, but only 0.7 percent of uranium ore contains fissile isotopes. If that isotope were purified further, it would produce about 2,000,000 times as much energy per mass, not counting the weight of the oxygen that reacts with the burning coal. On a spaceship where you must carry your oxygen as well as your fuel, the ratio is more like 10,000,000 to 1.

Another way to express this is that a single fission reaction of uranium 235 releases about 200 MeV of energy, while one reaction between carbon from fossil fuel and oxygen release 4 eV of energy.

For fusion, the energy released per mass is even greater. Hydrogen naturally occurs mostly as an isotope containing one proton and no neutrons. Deuterium is a hydrogen isotope with one neutron, and tritium is an isotope with two neutrons. If we fuse one deuterium atom with one tritium atom and produce one helium atom with two protons and two neutrons, we are left with an extra neutron and 17.6 MeV of energy. This is not as much energy as is released with fission, but the uranium fuel is much heavier than the deuterium/tritium fuel, which means that fusion releases more than ten times as much energy for the weight of the fuel as fission.

Okay, then—let's do it. Let's smack some hydrogen together and make lots of energy. There is one barrier in the way, however. It's the Coulomb force. Protons are positively charged, so they repel each other with electrostatic force. You have to squeeze them to about 10^{-15} m in order to get the nuclear force to take over and bind them together.

In order to get two hydrogen nuclei that close, they have to smack into each other at 20 million meters per second. In a substance, the speed at which individual atoms are moving is essentially its temperature. That velocity correlates to 5 billion degrees Celsius (that's hot). The core of the sun is only about 16 million degrees C, but that represents the average velocity of all atoms. There is a distribution curve, and at the high end, enough hydrogen atoms are zipping along fast enough to sustain nuclear fusion.

Heat also causes pressure, and all that hot hydrogen is pushing outward, wanting to fly apart. Stars stay together because of their intense gravity. The gravity and heat at the core of a star are enough to push hydrogen together strong enough to fuse into helium.

The question then becomes how we make that happen on Earth. We have successfully done it in hydrogen (thermonuclear) bombs. Hydrogen bombs generate the heat and pressure necessary to trigger fusion by exploding a fission bomb. The fusion explosion is uncontrolled and self-sustained, at least long enough to wreak destruction.

That process is not going to be useful if we want to harness the energy of nuclear fusion to safely generate loads of electricity. We need a way to create a controlled sustained fusion reaction, without blowing up or melting everything. This has been an intense research and development project for decades. Can we finally be getting close to an actual fusion power plant?

A History of Fusion

In 1920, British astrophysicist Arthur Eddington first proposed the notion that stars are fueled by nuclear fusion of hydrogen into helium, an idea later validated by theoretical physicists and astronomical observations. The first patent for a fusion reactor came twenty-six years later, in 1946 in the UK, filed by George Paget Thomson and Moses Blackman. They designed a "pinch" method whereby an electric current was run through a hot plasma of hydrogen, inducing a magnetic field that would "pinch" the plasma into a thin line, a process they hoped could be made powerful enough to produce fusion.

The next milestone in fusion technology would later prove to be a hoax—on March 24, 1951, President Juan Perón of Argentina announced that his country had produced sustained nuclear fusion of hydrogen. He said this breakthrough would be "transcendental for the future life." The media ran with this story, creating a sensation. It seemed believable to the public, coming soon after the development of nuclear fission technology. It turned out that the announcement was about a century ahead of schedule. Before the hoax was revealed, however, it had the effect of panicking the governments of the United States, USSR, France, Japan, and the UK, who determined that they had to pour money into fusion research or be left behind.

In that same year, just two months after the Argentinian hoax, physicist Lyman Spitzer, working at what became the Princeton Plasma Physics Laboratory, presented a workable design for a fusion reactor, called the "stellarator." The idea was to use powerful external magnetic fields to contain a plasma of hydrogen in a twisted doughnut shape, confining it to such a degree that the requisite heat and pressure are achieved to cause fusion.

The stellarator design dominated fusion research in the West for the next twenty years. Spitzer was able to demonstrate that the basic concept worked—you could contain plasma using magnetic fields of this design. However, the plasma was very leaky, and progress toward creating fusion was slow. Eventually Spitzer concluded that the stellarator would never work. Focus shifted toward a basic understanding of plasma physics, rather than specifically creating fusion.

The first controlled fusion in a laboratory was actually achieved in 1958, in the Scylla reactor using the theta pinch technique for magnetic confinement, which compresses the plasma into a filament. Although a huge milestone, this approach was abandoned soon after because it was clear it could not scale up to industrial use.

The next step forward was the tokamak design developed in the

1950s by Soviet physicists Igor Tamm and Andrei Sakharov. This was also a torus design with magnetic confinement of a plasma, but the geometry was tweaked to fix the limitations with the pinch design and stellarator. At first the claims of temperatures achieved were not believed by Spitzer and other Western scientists, so the Soviets invited UK scientists to make their own measurements. They confirmed the Soviet claims, which then led to a burst of research into tokamak reactors. It remains a leading contender for successful sustained fusion today.

By the late 1970s, there were dozens of prototype tokamak reactors around the world that could produce controlled fusion, inching toward that next big milestone, called "burning plasma." This occurs when heat from fusion contributes to more fusion. Burning plasma is a step toward ignition, in which fusion in the plasma is self-sustained, without requiring further external inputs of energy. Ignition is likely necessary to get to the true goal, in which fusion is not only created but also produces more energy than it takes to create in the first place. It takes a great deal of energy to produce the massive magnetic fields necessary to contain the hot plasma. The resulting fusion reactions generate a lot of heat, and that heat can be used to make steam, which then turns a turbine to create electricity (the same method as all modern power plants). Even if the process works, if it costs more energy than can be produced, the entire project is useless as far as energy production is concerned.

Over the next few decades, the development of high-temperature superconductors—materials that conduct electricity without resistance, and therefore could be used to make powerful electromagnets with less energy—was a game changer for magnetic confinement reactors and made it feasible to get to fusion producing excess energy.

At the same time, inertial confinement started to make an appearance. Instead of using magnetic fields to confine the hydrogen plasma, it would instead be physically squeezed. First proposed in 1960 by John Nuckolls, that same year the first functioning laser was produced, providing the means for inertial confinement. However, this method produces the same problem as magnetic confinement, as it takes a lot of energy to produce all those high-power lasers.

In 1965, the first such fusion reactor was built at Lawrence Livermore National Laboratory (LLNL), with 12 beams and a confinement chamber that was 20 cm in diameter. Mirrors are used to illuminate the target over the entire surface. The heat generated on the surface causes it to explode, compressing the inner material, and so this functions somewhat like a hydrogen bomb.

This is referred to as the direct method of inertial confinement. There is also an indirect method, which has several technical advantages in terms of stability but comes at a higher material expense. Lasers are used to warm an outer layer of heavy metal, like gold, which then heats up to the point that it generates X-rays that bombard the fuel to produce the heat.

Based on their research, the LLNL built the National Ignition Facility (NIF), a laser-based inertial confinement fusion reactor, in 1997. Their design uses 192 lasers and a bean-sized gold container called the hohlraum. This remains a cutting-edge facility for this design, and in 2014 the NIF reactor was able to generate more energy from fusion than the energy absorbed by the fuel itself. This is not quite net energy production (it released only 1 percent of total energy put in) because there are other losses of energy along the way, but it was a step closer. The NIF may also be the closest reactor to ignition in the world. In 2021, they achieved burning plasma, increasing the energy produced to 70 percent of the total energy input into the reaction. Getting this close to ignition served as an important proof of concept.

As of this writing, and despite the science and technology slowly and steadily improving, no tokamak reactor has achieved net energy production. The JET (Joint European Torus) tokamak reactor in the

UK holds the world record for energy created by fusion, producing 11 megawatts of power averaged over 5 seconds of sustained fusion in 2022, mostly in the form of released neutrons. JET is an experimental reactor and is being used to run experiments to help design a bigger reactor that may achieve ignition. The current consensus is that in order to meet this goal, we need bigger reactors, on the scale of a functioning power plant, not a lab-based prototype.

Toward that end, the ITER project (which initially stood for International Thermonuclear Energy Reactor before it was decided that the name would just refer to the Latin word for "the way") was formed as an international collaboration of thirty-five nations, including the United States, Russia, China, and several European countries. While the idea was born in 1985, construction just began in 2020. The ITER project hopes to be the first tokamak reactor to generate not only sustained fusion but also net energy, and to actually produce electricity for the grid. ITER has some competition, however, from newer designs such as MIT's SPARC tokamak reactor whose advanced superconductors and powerful magnets allow it to be much smaller.

Technological hurdles remain. For the magnetic confinement method, continued improvements in magnet design and superconducting material make economic fusion feasible, but it is also necessary to have a high degree of control over the flow of the plasma. For inertial confinement, the main technological factor is the power and efficiency of the lasers, which have also been improving but similar issues of control and engineering remain.

Engineers also need to work out how to siphon off heat in order to run turbines without interfering with the sustained fusion. The ITER project plans to surround the reactor with a wall of material that will absorb the neutrons released in the fusion reaction (which is where 80 percent of the energy produced in fusion goes). This will heat up the material, which can then be cooled with water, some of which will turn into steam to run traditional turbines to generate electricity.

The Future of Fusion

So, how close are we to controlled, sustained, net energy–producing fusion? Will the ITER win out over the NIF, or some other competitor? As always, we can break this down to several more specific questions. Will commercial fusion be technologically possible? Will it be practical? Will it be cost-effective?

As is often the case, the first question may be the easiest to answer. Fusion power has turned out to be much more technically difficult than was first assumed. Disappointment over the years has led to the common joke that fusion power is thirty years away and always will be. But the tremendous advances cannot be denied. The ITER, which is designed to produce 500 megawatts of power, should come online after 2025. A prototype commercial fusion power plant, DEMO, is scheduled for 2040. Optimistically, we may have working commercial fusion reactors producing net energy by the 2050s.

At this point, the only thing that can stop fusion would be if the ITER and NIF both fail and the appetite for more investment in this technology dries up. Given how far we have come and how close we seem to be, this is unlikely. It is far more probable that the technology itself will not be the limiting factor.

Practical considerations and economic viability are the real variables when it comes to fusion energy. As discussed in the previous chapter, we must think about what our entire energy infrastructure is going to look like when fusion reactors might be coming online. How much of our energy will be centralized versus decentralized? How robust will energy grids be? How much grid storage will exist, and how much need for baseload energy production will there be?

Fusion, by necessity (at least until Mr. Fusion is invented), is large, centralized baseload energy production. Its utility will therefore depend on how much that is needed in the future.

The cost-effectiveness of fusion also depends on multiple factors, only one of which is the cost of fusion itself. As the technology develops, the cost is likely to come down. Optimistic projections indicate that fusion power can be four times as cost-effective as fission, but this does not account for the fact that fission technology is also improving. We cannot compare fusion to today's alternatives; we must compare to what will be available in the future when fusion is available.

One possible limiting factor to fusion cost-effectiveness is the cost of tritium. Most fusion projects today (such as the recent JET experiment that has the record for energy production from fusion) fuse deuterium and tritium into helium. Deuterium is widely available, but tritium is not, mostly because it has a short half-life (12.3 years). This makes the cost of fusion energy dependent on our ability to develop a way to cheaply produce sufficient tritium. The ITER project plans on surrounding the fusion reactor with lithium, which will generate tritium when it is bombarded with neutrons. If this works, then the fusion reactor can create its own tritium, possibly tipping the scales in its favor.

One potential advantage of fusion power is that it is relatively clean. The only by-product of the energy production itself is helium, which is a useful element. There are no greenhouse gases and no long-lived nuclear waste to be stored. But how much will these benefits be valued? Burning coal is cheaper, if you don't count the environmental costs of mining, the health effects of pollution, the subsidies, and the externalized costs of climate change.

At this point, whether the second half of the twenty-first century will see significant production of fusion energy is a coin flip. We will likely get there technologically, but it is possible that other options, like solar and battery grid storage, may simply price fusion out of the market. Or there may be a place for continued large baseload production (the minimum amount of power an energy grid needs, usually met by large-scale production), and fusion will be seen as a superior option to fission or fossil fuels.

Fusion technology will likely continue to advance and benefit from advancements in material science, superconductivity, and advanced electronics. In the more distant future, therefore, smaller and more efficient fusion reactors will likely become possible. Fusion may turn out to be the best option for space travel, or for powering extraterrestrial stations where wind and solar are not practical or even possible.

Fusion power may also advance by fusing different materials. One promising option is fusing helium-3 and deuterium, which may actually be easier than using tritium and waste less energy in the process. The limiting factor here is that helium-3 is very volatile and escapes from the soil into the atmosphere, and therefore there is no significant supply of it on Earth. However, helium-3 may be abundant on the surface of the moon because it comes from the solar wind. It is blocked by Earth's atmosphere but not from hitting and building up on the lunar surface. Mining helium-3 may therefore not only fuel lunar stations but also become a lucrative lunar industry.

Because fusion energy has a higher specific energy than any other option, it is likely to play a role in our energy future in some form. The only potential form of energy that is even greater than fusion would be matter-antimatter annihilation. When this occurs, all the mass is converted to energy. The energy released is tremendous, but how much of this energy can be harnessed is another question.

The problem with antimatter-based energy is that there is no naturally occurring supply of antimatter, which takes a great deal of energy to create and to store (remember, it can't come into contact with regular matter). By one estimate, you get only a tenth of a billion (10^{-10}) of the invested energy back.

Any energy technology based on antimatter, or some other method of mass-energy conversion, cannot be extrapolated from current technology, or even our current knowledge of physics. Other, more exotic forms of energy production, like harnessing the immense gravity of black holes, are even more speculative and in the far distant future. Quixotic claims of zero-point energy or cold fusion remain in the realm of pseudoscience, as we will discuss in chapter 25, so I wouldn't hold my breath for those. These energy sources remain in the realm of science fiction, and all we can say at this point is that any such theoretical technology is indefinitely in the future. For now, fusion power is the best we can do.

It seems inevitable that fusion power will cross over the line to practical and economical mass energy production. The variable is how long it will take, but if not this century, then very likely the next. Once that line is crossed, fusion power will likely replace any other method of large, centralized energy production—fossil fuels, nuclear fission, hydroelectric, and others. As the technology progresses, getting smaller, cheaper, and more efficient, it has the potential of displacing even solar or wind power, at least on Earth, as it has a smaller land footprint. The sun is basically a giant fusion reactor in the sky spewing out free energy, but solar will be best for distributed and smaller power generation. Most of our power will likely come from fusion. This will especially be true as we move out into space.

Over the coming centuries and millennia, therefore, it is entirely plausible that fusion may win the long race over all competitors, producing massive amounts of clean energy with a small footprint. For the vast majority of future human history, fusion will likely rule, getting smaller, more powerful, more reliable, and more efficient over time. Fusion may be the technology that gets us into the Type I Civilization club.

17. Mature Nanotechnology

For good or ill, it will powerfully transform our world.

Nanotechnology is often thought of as the ultimate technology of the far future, one that could grant almost godlike powers, solve a myriad of problems, and supplant other technologies. Such wild claims should trigger a healthy skeptical reaction, but what exactly is nanotech, how plausible is it, and how far in the future is it likely to be realized?

Nanotechnology refers to any manipulation of matter at the nanoscale—a nanometer is one billionth of a meter. This is at the scale of molecules. A DNA strand is 2.5 nanometers (nm) in diameter. For scale, if a marble was 1 nm, then the entire Earth would be one meter in diameter.

The idea of manipulating matter at this scale is credited to physicist Richard Feynman, from his 1959 lecture, "There's Plenty of Room at the Bottom." But the term was coined by Japanese scientist Norio Taniguchi, who was specifically referring to semiconductor manufacturing. The idea of the potential of nanotechnology really became popular with a 1986 book called *Engines of Creation: The Coming Era of Nanotechnology* by MIT scientist K. Eric Drexler. In it, Drexler introduced the idea of nanomachines such as an assembler, that could make copies of itself in addition to performing other construction tasks.

Since then, the term "nanotechnology" has been somewhat distorted by companies looking to market their products as cutting edge and high-tech because they include nanotech. The common use was therefore expanded to include theoretically anything that has components at the nano scale. As we discussed in chapter 11 on material science, this can include a number of materials with nanoscale features or made with nanoparticles. It could also include things like computer chips, which legitimately have nanoscale features. In 2016, Lawrence Berkeley National Labs created transistors that were only 1 nm in size.

The "made with nanotechnology" stamp became as common as the marketing term "space-age technology." Everything created after Sputnik was launched in 1957 is technically "space-age." Nanotech enthusiasts therefore created the term "molecular nanotechnology" or MNT, to refer to nano-sized machines, not just objects with nanoscale features.

At present, MNT does not really exist. There is early research trying to work out some of the basics of machines that can function at the nanoscale—tiny motors, gears, and levers—but no current applications.

Plausibility of MNT

Nanotechnology may be one of those things that sounds great on paper, but simply doesn't work in the real world. Do we have any early success or real-world examples that demonstrate nanotech proof of concept?

Some enthusiasts point to life as an example. One could argue that at the subcellular level, at the size of organelles that make cells function, we have nanoscale machines. Proteins that unzip DNA, convert messenger RNA into proteins, or act as cell-surface pores are all nanoscale machines. In fact, some scientists believe that if we want to build nanomachines we should use the infrastructure of life. Why build a machine when you can program DNA to make it? This is an excellent point, but it is more a justification for nanotechnology and biotechnology to progress together, rather than to pick one over the other.

What the life example does, however, is show that the laws of physics allow for nanoscale machines to operate. The laws of physics function differently at widely different scales. This is why a fly can walk on water when a large crab can't—water tension is a much greater factor relative to the fly scale than the crab scale. As we begin to design and build nanomachines, based upon our macroscale understanding, we may find that things behave differently at the nanoscale.

One example of this is Brownian motion, the random motion of small particles in a fluid (like motes of dust dancing in the air) from being bombarded in random directions from molecules of the fluid. We don't have to worry about Brownian motion at the macroscopic scale, but it is critical at the nanoscale.

Wonky behavior of objects at the nanoscale, however, is likely to be a hurdle rather than a full stop. We will need to master how machines behave at this scale, which will likely mean it takes longer to develop molecular nanotechnology than early optimism would indicate. But these issues do not seem unsolvable.

There is also a real-world proof of concept, even if very primitive (for nanotechnology). In 1988, IBM's Zürich Research Institute spelled the letters "IBM" with thirty-five xenon atoms using a scanning tunneling microscope, showing atomic-level control is possible. There is still a question, though, of whether we can control machines at the nanoscale well enough to do what we want. How do we get the tiny claw to pick up the molecule, move it where we want it, and then assemble it with the other molecules? Further, how do we keep nanomachines from doing what we don't want them to do?

There are a few options here. One is to design nanomachines so that they follow a simple repetitive function, but in the aggregate that function produces a desired macroscopic effect. This would require insect-level intelligence. How do bees build a complex nest? No bee guides the process. Each bee just follows a simple algorithm that produces a honeycomb when repeated over and over. This approach is great for building materials or simple shapes or designs, but by itself would not be able to produce complex machines. Of course, there could be a series of nanomachines, each doing their part in a sequential process that creates the final design.

Another method is to have larger nanomachines, perhaps in the 100 nm range, big enough to contain some actual computing power. These machines could enact far more complex instructions and may even be able to function as shepherds for smaller nanomachines. It would be an ecosystem of different nanomachines of differing sizes, functioning together to complete a complex process.

Lastly, we could also employ a form of external control, such as moving nanomachines with magnetic fields or lasers. The fields could be coordinated by macroscopic supercomputers, with the nanobots as passive drones.

We may find that a combination of all these approaches is best. Controlling nanomachines is therefore challenging, but not impossible.

Potential Applications—Manufacturing

All right, we eventually figure out how to make a variety of nanomachines and to precisely control them. What does this mean? William Powell, who was the lead nanotechnologist at NASA's Goddard Space Flight Center until his untimely death at age thirty-six, famously said of this process, "Nanotechnology is manufacturing with atoms."

Some experts have referred to this process as "mechanosynthesis" to distinguish it from chemosynthesis. In the latter, molecules are attached by encountering each other through random thermal motion. In mechanosynthesis, molecules are precisely attached at specific receptor sites. In the end, the molecules are bonded to each other by the same chemical forces, but mechanosynthesis allows for atomic-level precision.

The obvious application for advanced MNT, therefore, is fabrication. Raw materials could be directly converted into either advanced materials (e.g., fabrics, metals, plastics, resins) or finished products. The resulting goods could theoretically have atomic-level perfection and could even have internal structure not possible with traditional manufacturing techniques. This could potentially solve many problems, such as creating error-free graphene.

As we extrapolate out nanomanufacturing, we eventually get to the point that all you need are raw materials, nanomachines, an energy source, and designs. With these four things, you can build a city from the ground up. Garbage will just become raw material, a source of atoms and molecules, and anything can be recycled.

First there will likely be nanofactories, centralized production using expensive nanomachines and systems. Nanofacturing will probably not immediately replace other forms of manufacturing. Rather, they will supplement more traditional methods and develop alongside advanced 3D additive manufacturing and CNC machining. Each method will be used where it is optimal, with factories even combining all of them.

Very advanced nanofacturing, however, has the potential to replace all other methods. This will look like the replicators of *Star Trek*—just put in a monocolored card containing the instructions, and the desired object appears. (In *Star Trek*, the replicator also uses transporter technology, but we'll ignore that for now.)

Nanofacturing can also be decentralized, until every home has one as a standard appliance. There will then no longer ever be the need to buy anything, except raw material and designs. As with 3D printing, many designs will likely be open source, but if you want the latest gadget or a proprietary work of art, you rent the right to reproduce one of them (or however many you need). Likely for a long time, perhaps indefinitely, handmade objects will still be in demand. At some point, however, the replication will be so good you won't be able to tell the difference.

Eventually we may have a society with no garbage, no waste, and manufacturing so easy that you can endlessly re-create your environment, furniture, and appliances to suit your mood. This type of technology will make the physical world more and more digital in character—as easy to change as altering a picture on a computer screen. It will also be highly disruptive to our current economic reality.

Science fiction author Valis Umbra created a nanotech future in which he imagined:

> It will change everything. We won't need governments and corporations. The very concepts our economic and social systems are based on: money, wealth, privilege, will be meaningless. What use is money when everyone can build everything they need?

Medicine

Nanotechnology also has potentially profound implications for medicine and health. Cells operate on the nanoscale, and the ability to design medical instruments that function on the cellular scale would usher in a new age of medicine. We can imagine some simple, yet powerful, applications, such as tiny nanomachines crawling along your arteries and removing any cholesterol plaque buildup, not too far in the future.

Nanosurgery with progressively smaller instruments is also likely to be an early application. At first these might use conventional surgical methods, like suturing and cautery, but at a tiny scale with absolute precision. There would also be no need to make any incisions. Already, surgery using small cameras has reduced the need for large incisions, allowing for less traumatic surgery with much faster recovery times. What if the same result could be accomplished with just an injection of nanosurgical machines through a needle? Think *Fantastic Voyage*, but without having people inside a tiny submarine, just tiny machines.

These nanomedical bots could include a variety of tools like cameras to see where they are and visualize anatomy, even creating 3D scans that are beamed to external computers that control the nanosurgery. With this level of fine control, new possibilities, such as removing cancerous tumors in such a way that leaves no cancer cells behind, are created.

Nanorepair at the cellular level is the stage beyond this, replacing macroscopic (called "gross" in medical parlance) surgery with microscopic repair. This could precisely and securely reattach tendons, repair tears in connective tissue, or repair blood vessels or microscopic damage in organs. Past this even is subcellular repair, mending individual cells by clearing out the buildup of toxic proteins, fixing malfunctioning organelles, and altering the function of cell-surface proteins. Finally, at the molecular level, nanomachines could repair DNA, elongate telomeres (the ends of chromosomes that shorten with age), and undo much of the cellular damage that occurs with disease or aging.

Cellular and molecular-level nanomedicine has the potential to not just treat but to also rejuvenate or even regenerate—grow missing limbs or replacement organs. At the very least, nanomedicine has the potential to cure many currently incurable diseases and disorders. It can make surgery much less invasive and traumatic, and more precise and successful. It can repair damage to keep people high functioning and healthy for much longer and potentially extend life.

The last bit remains controversial—even with mature nanomedicine, what is the ultimate potential for extended human life spans? All systems tend to break down, and it becomes mathematically untenable to prevent this for longer and longer time periods. There is also the matter of the brain. You cannot replace the brain, as that would be the same as death. Brain health is likely to be the ultimate limiting factor on life extension, bringing us back to the brain-machine interface and the technology to migrate our neurological function to machines.

Programmable Matter

Nanomachines can not only be used to make stuff but can also be the stuff itself. Some nanoscientists have proposed ideas for advanced programmable matter made of tiny nanomachines, sometimes called "foglets" (because they form a fog of tiny machines). These would be in the somewhat larger 100 nm range, and designed not to alter matter to create objects but to connect with each other to form those objects.

Imagine building a house. The land is prepared and perhaps a solid foundation poured. Once the foundation is ready, the appropriate amount of programmable matter is delivered and dumped onto the foundation. You then use some controlling interface (perhaps virtual reality) to command the foglets to form into a house, including every bit of furniture, fixtures, artwork, appliances, and so on. The resulting home could then be altered at will.

I won't fall into the futurism fallacy here of assuming we will use advanced technology only to replicate existing functionality, like a home. A foglet-based domicile could take on new forms and functions not possible with conventional technology. Perhaps such homes, for example, will not have doors—a wall will just create an opening when and where necessary. They could have interactive functionality, guiding guests to a desired location, managing access for pets, trapping would-be intruders, providing first aid, to name a few.

The same might be true of a vehicle, although really, it's not a vehicle but rather a backpack or piece of luggage. The pack is made from foglets that are carried and can be commanded to form into anything a person might need. It is every tool, every device, every personalized form of transportation. Imagine George Jetson folding up his flying car into a briefcase to carry with him. This may not seem so impossible someday.

Clothing might be similarly programmable, changing any property as desired—color, texture, thickness, protection, insulation, and of course style. It can function as armor, as a wet suit, a parachute, or a breathing mask as necessary. It might even be a space suit, which will be handy for those living in settlements on the moon or Mars.

You might need to top off your supply of foglets from time to time, or upgrade them, but otherwise it could theoretically be the only stuff you need. It's likely that programmable matter cannot become anything, like complex machines, computers, batteries, and so on. Therefore, there might be individual modules with specific functions to integrate with the programmable matter. Of course, they may all be nanofactured, even the programmable matter.

The Risks of Nanotech

Like most technological advances, new capabilities have risks in addition to benefits. As our technology progresses to extreme levels, those risks also become extreme. The primary risk of nanobots, imagined by science fiction writers and scientists, is that they will get out of control.

One potentially apocalyptic error could occur with nanoassemblers that make copies of themselves. Safeguards can be put in place to limit the number of copy generations, or their density or total number, but these safeguards might be bypassed by fortuitous (for the nanomachine) errors in copying, inadvertently creating new abilities.

This is sometimes referred to as the "gray goo scenario." A self-replicating nanobot goes out of control, with geometric increases in the copy numbers (e.g., doubling the number of nanobots every minute), until eventually the entire surface of Earth is transformed into a uniform gray goo of nanobots.

A similar scenario might unfold if there are already many nanobots and their instructions to manufacture objects somehow break out of the imposed limits. In this case, future aliens visiting Earth might find the entire surface is covered in paperclips to a depth of 20 meters.

A realistic concern is nanowarfare—nanomachines specifically created to destroy enemy infrastructure, and perhaps even enemy combatants. It does not take much imagination to see how this can go horribly wrong. It's something we should probably just agree not to do, like building Cylons. Just don't go there. Still, abuses and nefarious uses are easy to imagine. Valis Umbra again:

> If the next generation of nanotechnology could be used in medicine to repair organs and tissue from inside the human body, then it could just as easily be programmed to destroy them, making it the ultimate weapon of assassination.
>
> Imagine clouds of these things flying to their targets to either be breathed in like a virus or ingested with food or drink, and then creating fatal haemorrhages or lesions that lead to death from apparently natural causes.

The Even Farther Future

Assuming we survive the nanopocalypse, where might this technology go given thousands or even millions of years of perfection? We can speculate about ever-smaller machines with ever-finer control over matter. Beyond nanotech there is picotech, which can manipulate matter at the scale of protons and neutrons. Beyond that is femtotech to manipulate matter at the subatomic level.

The ultimate expression of this thought experiment is Plancktech—manipulating not only matter but also space-time itself at the fundamental scale of existence. The Planck length is like the pixel size of reality; it is the ultimate graininess of existence. Manipulating reality at the Planck scale, if it is even theoretically possible, would grant almost limitless power and control over existence.

We can't really conceptualize this level of technology, except in the abstract, but perhaps there are alien civilizations out there, millions of years more advanced than humans, operating at something like this level. Let's hope they're friendly.

18. Synthetic Life

The lines between biology and machines are being increasingly blurred.

The ultimate test of our knowledge and technology's power to manipulate life is the ability to not just alter existing life, but to create completely synthetic life. Imagine designing an entirely new organism from the top down and then building it from the ground up, unconstrained by prior evolution. This would be a powerful platform from which to spawn countless applications.

While we have not yet accomplished this goal, the field is advancing toward it, and there doesn't appear to be any theoretical hurdles in the way. This is a future technology we may achieve simply by continued linear incremental advances. Once the basic technology is in place, it's easy to let your imagination run wild with possibilities.

In 2010, American geneticist Craig Venter announced that his team had created a completely artificial cell. After ten years of research at a cost of $40 million, they were able to create from scratch a minimal artificial genome and then place it in a membrane to construct their artificial cell.

The goal of this research is to build a single living cell with minimal functions. That minimal genome can then be the foundation on which to build more and more complex life-forms. That in itself is an interesting biological question—which genes are necessary for life?

In 2016, Clyde Hutchinson and colleagues published a study in which they started with a bacterium, *Mycoplasma mycoides*, and then tried to strip it down to the lowest number of genes necessary to keep it alive. They were able to reduce it to a mere 473 genes. For perspective, humans have about 20,000 genes, and free-living bacteria have 1,500 to 7,500 genes. Of these 473 minimally required genes, 149 still have an unknown function.

Either by stripping down or building up individual cells, scientists are getting relatively close to having a platform that represents the fundamental components of a living cell. We can already manufacture DNA. The ultimate goal is to have a catalogue of genes or suites of genes with known functions that can be selected and then assembled into a synthetic genome. That genome can then be placed in blank cells that lack genetic instructions of their own but contain the basic machinery for assembling proteins from the implanted genes, resulting in a functional artificial cell.

Even that alone would be potentially useful. Since such cells were designed and assembled by scientists, their properties and functions could be theoretically completely understood, as much as any machine. Perhaps a better analogy would be a piece of software, for that's what genes are in essence. Building an artificial genome might be akin to programming a complex application.

Bacteria-like artificial cells could serve a variety of functions, such as converting waste, cleaning up toxins or spills, and supporting agriculture by fixing nitrogen or other nutrients. Industrial applications are also likely, such as chemical or pharmaceutical manufacturing. Brewing, cheesemaking, and any food process currently performed by yeast could potentially be done by artificial life with designer properties.

Because of safety concerns, medical applications would likely come later. However, theoretically we could develop cells that could treat diseases, function as another class of therapeutics, or be a drug-delivery system. They could replace or augment our natural flora (the bacteria that line all our membranes) in order to enhance our biological function and protect from infection. Specific cells that are safe for the host but able to hunt and kill cancer cells would be another potential groundbreaking application.

Getting to a basic single artificial cell is still likely a few decades off, and any of the single-cell applications just mentioned may be a few more. I suspect that significant impact from uses of artificial life is something we will be seeing toward the end of this century. It's possible, however, we may run into unanticipated hurdles or safety concerns that will further delay real-world applications.

Artificial multicellular life will take longer still. It would likely be easier to develop such organisms rather than build them directly. In other words, we program an artificial embryo and then let it develop into a mature organism. It is also possible we could make the individual cells, then 3D print them into an organism, but this would probably be more difficult. Perhaps the use of one method or the other will depend on the needs of the different applications.

It is harder to imagine what roll artificial multicellular organisms would have, as this technology is potentially more disruptive and dangerous. Such creatures could be designed as food, or as pets, or perhaps as workers or servants.

Such applications raise an important question—why would we make artificial life when we could evolve or genetically engineer conventional life? Theoretically, the reason would be to have a greater level of control and more open-ended possibilities. These would be essentially competing technologies, with their own strengths and weaknesses, although artificial multicellular life is too far in the future for us to predict what those will be.

With genetic engineering of existing life, we have the benefit of starting with organisms that are the result of billions of years of evolution on the cellular level, and hundreds of millions of years of evolution of multicellular anatomy and function. This is a lot of trial and error, with enough time to work out many of the kinks.

Although, future narrow AI could likely simulate all that trial and error in no time.

Further still, evolved systems are also messy and can be inefficient. The majority of human DNA, for example, is "junk" and not necessary for normal function. Evolution works with what it has and comes up with good-enough solutions that are often suboptimal. Designing an organism from the ground up can eliminate all this constraining history and baggage.

Consider the vertebrate eye, for example. There are many aspects of its anatomy and function that you would never design on purpose. The most famous example is that the light-sensing cells are on the bottom of the retina, beneath other cell layers, including the blood vessels and nerves. It would make more sense to put the light-sensing cells on the top layer, unobstructed and without the need for a blind spot where the nerve fibers go back to form the optic nerve.

One question that keeps coming up is whether we will be able to change whatever we want with sufficient genetic engineering technology or if it will be easier at some point just to start from scratch and build from the ground up. Usually, the answer to such technological questions is that both will be true in different contexts, so these competing technologies may coexist as complementary.

There are a couple of areas where synthetic life may have an advantage, however. One is in extreme conditions. Even though "extremophiles" already exist, these are mostly bacteria or archaea—single-celled creatures. Multicellular extremophiles able to live in extremely harsh conditions are rare. What if we need organisms that can survive in space or are adapted to life on the surface of Mars or Ganymede, or the oceans of Europa? We may run up against the limitations of existing biology, even with extensive engineering, and need specific functions that don't currently exist in the natural world.

Another feature we could engineer with artificial life but is not possible with mere genetic engineering requires fundamental changes to the building blocks of life itself. For example, what if we wanted to change how DNA itself works? DNA is a code built from four components, the four bases adenine, guanine, thymine, and cytosine (AGTC). This creates a four-letter alphabet, with three-letter words that code for twenty different amino acids in addition to codes for where to start and stop. These four letters also exist in pairs, with A and T binding together and G and C binding together, allowing for DNA to be a double helix with two complementary strands. Everywhere there is a G in one strand, there is a C in the other. DNA can replicate by splitting into two single strands, then assembling complementary bases to form two double-helix DNA.

Scientists, however, have already developed a form of DNA with six bases forming three base pairs instead of two. They used the triphosphates d5SICS and dNaM, which they designated X and Y in the DNA alphabet. The scientists also demonstrated that these new base pairs were able to incorporate into the DNA structure and were able to function with the normal cellular machinery, even replicating. Such DNA could code information more densely and accommodate a more complex alphabet, meaning that it could control the placement of more than the standard twenty amino acids. This could allow for proteins and therefore biological structures and functions not possible with evolved biology.

Speaking of the standard twenty amino acids that all known proteins are based on, it is an interesting question in and of itself as to why only those twenty amino acids became part of life. There are many more possible, but life settled on only twenty out of which to build all proteins, and therefore include in the genetic code. There are partial explanations for this having to do with their chemical properties, such as how they appear to bind more competitively than nonprotein amino acids. Still, there is no hard limit to these twenty amino acids, and expanding the repertoire would allow fascinating possibilities.

Whether or not these specific changes will turn out to be useful is neither here nor there. Rather, it is possible that biologists may hit upon new forms of DNA or proteins that are not feasible with existing life and will require entirely synthetic life.

What Are the Risks?

There are, of course, risks to this technology that increase as the technology matures. One is easy proliferation. Imagine, as some science fiction writers already have, if synthetic life technology becomes so automated it is possible to have a desktop appliance that can crank out artificial organisms. This could be run with an application similar to Photoshop, but for life. AI will take care of all the biological details. You just tell the software interface to put an arm there, and the AI will put together the genes necessary to build an arm. How many fingers would you like? How big do you want the claws, or perhaps just soft finger pads?

It is incredibly likely that synthetic life technology will be highly regulated, but the cheaper and easier it becomes, the more difficult it will be to regulate. Safeties may be built into the software, but those could also possibly be hacked. We can imagine safety precautions like in *Jurassic Park*. Remember the lysine contingency? All the dinosaurs were genetically unable to make lysine, so they would die without a special supplement. We see how well that worked out.

The *Jurassic Park* example begs the question of how quickly synthetic organisms would be able to evolve, if at all. Their genomes could be designed to minimize mutations, for example. For that matter, synthetic organisms could all lack any ability to replicate. Developing such capability itself may be illegal. Therefore, even if the basic application is widely available, it may be as difficult as obtaining weapons-grade plutonium to acquire the software necessary to design organisms capable of any replication. No replication, no evolution, problem solved. (This also would prevent patent infringement by keeping people from copying proprietary organisms.)

There is, of course, no full-proof safeguards. Anything can be hacked. But there are commonsense methods to limit the probability of a synthetic life apocalypse. In addition to the inability to reproduce, they may not be able to consume normal food. They might need synthetic food compatible with their biology.

There would be other sources of risk in addition to home-brewed synthetic life. Industrial production might be hacked or hijacked, producing large volumes of single-celled life or even multicellular life that could wreak havoc before they are contained or reach the limit of their life span.

Like many of the future technologies we have been discussing, if deliberately used for military or terrorist purposes, advanced artificial life technology can be quite scary. We might imagine an *Aliens* scenario—a "perfect organism" designed from the ground up to be supersoldiers. Such creatures could be deployed in enemy territory, subduing the local population before dying themselves, leaving the way open for conquest. This is very reminiscent of the robot soldier scenario. Perhaps future synthetic soldiers will have to fight it out with future robot soldiers to see which technology has the privilege of wiping out humanity.

Far Future

Assuming we survive the synthetic life apocalypse (and all the other apocalypse scenarios we have mentioned), what might the far future possibilities be for this technology? One advantage of biological technology over machine technology is that biology can be grown. Once you have the design for a synthetic biological machine, you only have to provide a suitable environment and adequate nutrients to produce as many such machines as you need. Biology, in essence, is an effective self-assembling and self-replicating manufacturing platform.

In the science fiction show *Babylon 5*, for example, the Vorlon spaceships were actually giant living creatures. That is some extreme biological technology, but it highlights the possibilities once we have total control of biology at every level.

We may also explore what is called "semisynthetic life," which is essentially a hybrid where significant synthetic components are added to a natural organism. For example, we could make entirely synthetic organs to replace or augment our existing ones. Synthetic organs can perform tasks not possible with standard biology, such as extreme detoxification or superimmunity. Synthetic muscle could provide preternatural strength, or synthetic filaments could reinforce our bones and tendons. Synthetic biological eyes could see in an enhanced spectrum or with higher resolution.

At the subcellular level, synthetic mitochondria could provide extremely efficient energy or burn alternative fuel. DNA repair organelles could slow the aging process and eliminate most cancers.

There are even more extreme possibilities, which shows how many paths future technology can take us down. We may develop synthetic symbionts—creatures designed to merge with a human host, like the Trills on *Star Trek: Deep Space Nine*. Instead of (or in addition to) implanting computer chip–style AI, we could augment our cognitive abilities by merging with a superintelligent synthetic brain. This brain could also live independently for centuries, and when our current body dies, it could disconnect (taking our memories and personality with it) and be transferred to another host. This may be another path to extreme life extension.

This gets to a concept we have previously touched upon. How much of our future technology will be biologically based? Will we build robots or artificial organisms, perhaps even artificial intelligent life? To what extent will we enhance ourselves with genetic engineering versus synthetic components versus computer-based cyborg components? Will we even grow our computers from biological components, using some form of DNA to store information and carry out information processing?

The easy answer is that we will do everything, but the unknown factor is what the balance will be, and for which specific applications will biological rather than machine-based technology work best.

There is something to be said for the ability of biological systems to self-assemble. We could theoretically terraform an entire world by just dropping seeds that will grow and create not only an entire ecosystem, but also an entire foundation for a civilization. We will cultivate our habitats, vehicles, energy sources, food, and other resources.

At the same time, the lines between biology and machines are being increasingly blurred. Future nanotechnology may also be able to self-assemble and do everything listed above by itself. Biology, in essence, is just one type of machine technology—organisms are just biological machines. So, it is possible that the distinction between biology and machine technology will not only blur in the far future but will also be entirely erased.

In fact, the lines between natural and artificial are also likely to dissolve. What is "natural" is already difficult to precisely define. Why are things that already exist in nature privileged in our minds? They are mostly the quirky outcomes of evolutionary forces, which could just as easily have produced an entirely different set of results. Humans are now part of those evolutionary forces too, and it is in our nature to use our technology to change whatever aspects of our environment suits our needs.

As we have seen in many chapters in this section, a major theme of technological advance, especially disruptive technologies, is that they blur existing lines and categories. The digital and analog worlds are blending as we overlay the physical world with digital information, and build our physical infrastructure digitally. We may merge with and partly become our technology. Life itself may become entirely synthetic.

We may eventually live in a world where there are no such distinctions. Everything will be digital information. We will be digital information. Distinctions such as natural, artificial, machine, and biological will seem hopelessly quaint.

19. Room-Temperature Superconductors

If we get there, it will supercharge our technology.

If you are old enough to remember the 1980s and had any interest in science and technology at the time, you likely remember the breathless reporting about the high-temperature superconductivity breakthrough. Every popular science magazine covered the story, usually with a picture on the cover of some object levitating above a superconducting magnet. If the popular reporting was to be believed, then superconducting material was about to change the world. Four decades later, we're still waiting.

Despite this false hype, true room-temperature superconductors would be a game-changer technology. You also have to love any technology that has the term "super" in the name. A superconductor, as the name implies, conducts electricity extremely well. In fact, it conducts perfectly, with zero resistance, meaning no energy is lost to heat or any other waste. In an electrified world, the benefit of superconducting material should be obvious. Every electronic device made with such material would be more efficient, use less energy, and would not have to deal with waste heat. If such a material were available, it would transform our technological world.

Such materials exist, but the title of this chapter also includes a critical caveat—what we are really talking about are materials that are superconductors at room temperature (or, ideally, at whatever temperature they might experience in normal use). There are other caveats as well. An ideal superconductor would have this property at normal pressures, or even in a vacuum. It would also be nice if the material were strong and ductile, or at least not brittle. Lastly, being reasonably cost-effective is important in terms of possible applications.

The Physics of Superconductivity

The phenomenon of superconductivity was discovered in 1911 by Dutch physicist Heike Kamerlingh Onnes. Onnes won the Nobel Prize in physics in 1913 for his work with extremely low temperatures, being the first person to cool helium into a liquid, which happens at just 4.16 kelvin (K) above absolute zero. This was critical to his discovery of superconductivity because at that temperature, the element mercury becomes a superconductor.

Most conductive materials will experience decreased resistance to conduction as their temperature decreases. However, superconducting materials, like mercury, have a critical temperature at which their resistance rapidly decreases to zero. Mercury being a superconductor at 4.16 K is not very practical, but it did allow for further research into superconductivity itself (especially since liquid helium could be used to cool substances to that level). Liquid helium is expensive, however, limiting the amount of research.

Further study quickly resulted in the discovery in 1912 that tin had a superconductivity temperature of 3.8 K and lead, 7.2 K. These materials could be made into wires and were much easier to deal with than mercury, progressing the research. In fact, many elements (about half those on the periodic table) are superconductors, but all with extremely low transition points (colder than that for lead).

It turns out there are two types of superconductors. Type I (like tin, lead, aluminum, titanium, and others) are superconducting at normal pressure and extremely low temperatures. In addition to losing all resistance at their transition point, they also become completely resistant to magnetic fields. Another way to state this is that they have perfect diamagnetism, which is the ability to repel magnetic fields. Type II superconductors are all compounds (no simple elements) that have higher transition points.

In 1957, John Bardeen, Leon Cooper, and Robert Schrieffer proposed the BCS theory for how superconductivity works. A simplified explanation, without getting into the technical details, is that in a superconducting state, electrons form into Cooper pairs. These pairs move through the lattice of the material together. When the first electron moves through the lattice of positive atoms, it pulls the lattice together a bit, forming an area of greater positive charge. This positive charge attracts the second electron in the pair, pulling it forward. This also explains why temperature is such a critical feature. Temperature is a measure of how much atoms are moving on average. When the temperature gets too high, the atoms in the lattice are jiggling too much for this delicate configuration to be maintained.

The first type II superconductor discovered in 1986 was cuprate-perovskite ceramic materials with a critical temperature above 90 K. This is still very cold: 90 K is –183°C. But this temperature crosses a significant threshold—77 K, the temperature at which nitrogen is a liquid. Liquid nitrogen is much cheaper and more available than liquid helium, which is a game changer for superconductivity research and applications. Superconductors above the liquid temperature of nitrogen are often referred to as "high-temperature superconductors."

This was the discovery that led to all the popular hype mentioned above. It was a big deal. The assumption that because the record for high-temperature superconductivity jumped so much, such advance would continue and we would soon be up to room-temperature superconductivity (the futurism fallacy of projecting current trends into the future), was a mistake. That's not what happened. It has proven very difficult to get anywhere near room temperature.

Cuprates continue to be the record setters for high-temperature superconductors. The current record is held by a cuprate of Hg–Ba–Ca with a critical temperature of 134 K (–139°C).

Physicist Elliot Snider and his team broke the record in 2020 with a maximum transition point of 287 K (13°C). You probably have guessed, however, that there is some important caveat to this claim, and you would be correct. The material is only a superconductor at this temperature while under extreme pressure: 220 gigapascals, which is over 2 million atmospheres of pressure.

While this is scientifically interesting, and may further our fundamental understanding of superconductivity, such extreme pressures mean there is no direct application for this class of material. It's also not likely a route to practical room-temperature superconductivity. The cuprates might not be either.

I don't want to downplay the significance of high-temperature superconductors, however. The ability to be cooled with liquid nitrogen is practical for many applications, such as the extremely powerful magnets that are necessary for fusion reactors and for particle colliders. But this requirement does significantly limit these potential applications.

Right now, we do not have a candidate material that will have all the necessary properties to be a practical superconductor at typical room temperature, the kind you would need for everyday electronics, an electric vehicle, or the power grid. When are we likely to cross this threshold? It is impossible to predict. This goal will not result from any incremental advances or extrapolation of current technology. It will require a genuine breakthrough—the discovery of a new type of material with the necessary properties. This can happen in ten years, or a hundred years. Perhaps we will find that such a material simply does not exist.

Let's fantasize anyway about what such a material would mean for future technology.

A Future with Superconductivity

If we're going to speculate, let's go all the way. Let's say at some point in the not-too-distant future, scientists discover and then perfect a compound, which we will call "conductium," that is a strong and ductile material, easy to shape and use as a wire or as part of a circuit board, and has a superconducting critical point at 50°C, which is enough to cover most human environments. Conductium is made of raw materials common enough to be mass produced.

While we're at it, there are a couple of other properties that would be nice to have: high critical current density and critical field. The critical current density determines the maximum amount of electricity that can be passed through the material before superconductivity stops, while the critical field is the maximum magnetic field the material can tolerate.

With conductium readily available at the local hardware or electronics store, pretty much anything that conducts electricity will be made at least partly of this material. In 2021, it was estimated that about 2 percent of global electricity demand was needed to run the internet. If we include all information communication technology, the figure is between 5 and 9 percent. About 10 to 20 percent of a computer's energy is lost to waste heat. Then further energy must be expended to cool the computer, or else the circuits would be fried. There are estimates that this figure can rise to 20 percent by 2030, although the precise number is difficult to nail down because of increasing efficiencies and changing use. Regardless of the specific percentages, a great deal of power is used by computing devices, and this is likely to increase.

In addition, our energy grid would save about 7 percent of lost energy if it were constructed of conductium. It may even be possible to make energy storage devices that are simply circuits with electrical current with zero resistance, so they don't lose energy over time.

Conductium opens up other new technological possibilities.

Computer chips that do not generate waste heat would be ideal for brain-machine interfaces. Powerful electromagnets could make technologies like MRI scans smaller and cheaper.

Maglev trains that travel on conductium rails would not need to be cooled, and therefore would be much more cost-effective. All types of electric vehicles would also be more plausible. Lighter and more effective motors and batteries would make electric flying cars, and electric planes, more efficient with greater range and reduced energy costs.

While fusion reactors are possible with liquid nitrogen cooled superconducting coils, a room-temperature version would be more powerful and more cost-effective, making it easier to cross the line to practical fusion.

Between energy savings, greater potential for miniaturization, and new electronic applications, the impact of a practical room-temperature superconductor would be significant and wide-ranging. Although not an application unto itself, it would facilitate many electronic devices and make new ones viable.

Without it, we could still eventually achieve all of the other technologies we discuss in the book, including fusion, quantum computers, brain-machine interface, and flying cars, but they would all be easier with a practical room-temperature superconductor. This is a reverse "steampunk" issue. Without room temperature superconductors, we can still steampunk our way to these technologies using existing materials. But if we did develop such superconductors, it would be like transitioning from steam to electronic technology—a significant game changer.

20. Space Elevators

Can such an astounding feat of engineering even work?

Imagine a giant cable attached to Earth at the equator. The cable rises into the sky, through the clouds and out of view. On a clear night, however, you can follow lights, some moving slowly up and down, on the cable as far up as you can see, until it eventually disappears in the distance above your head. If you could follow the cable all the way, you would see it climb 35,786 km (22,236 miles) up, where it connects to a space station in geostationary orbit. The station itself can be seen from the ground with a good pair of binoculars.

This certainly sounds like futuristic technology, and "space elevators" do make an appearance in many science fiction stories, but why would we want to build a space elevator?

The idea of a space elevator goes all the way back to Konstantin Tsiolkovsky in 1895. He published the idea of building an "orbital tower," designed like the pyramids, bigger at the base and then tapering as it became higher. The structure would use compression strength to achieve height. In 1960, the first proposal of a modern concept of a space elevator with a geostationary satellite connected to the ground by a long cable was proposed by Russian engineer Yuri Artsutinov. The cable design is considered a tensile structure because it relies on the tensile strength of the cable.

Fifteen years later, in 1975, American engineer Jerome Pearson published the first technical treatment of the concept of a space elevator. He remained a leading proponent and expert on the concept until his death in 2021. He also fleshed out an idea for a moon space elevator and explored potential applications. Pearson's technical details were even used by Arthur C. Clarke in his 1979 book, *The Fountains of Paradise.* Between Pearson and Clarke, space elevators entered the aerospace and public consciousness.

How Would It Work?

The basic structure of a space elevator is to have a base station on the ground where a long cable will attach. The station would need to be on the equator, with the cable going straight up to geostationary orbit. The cable would attach to the geostationary anchor, which would be in a perfectly circular orbit above the equator. A circular orbit is necessary because any eccentricity in the orbit would mean the cable would have to change length at different points in the orbit.

It is possible to have a base station not on the equator, but then it would have to be paired with another base station the same distance on the other side of the equator. Cables from these two attachment points would then come together to form a single cable that continues up to the station. The two stations essentially act as if they were one connection point on the equator, forcing balance.

The key to the engineering of the space elevator is the forces on the cable. Gravity is always pulling down on any cable. The longer the cable, the greater the total force of that gravity. However, if the cable is attached at the top to something in geostationary orbit, the cable will remain pointing out from Earth, so that as Earth rotates, the cable will also spin around, creating a centrifugal force pulling on the cable. Centrifugal force is an inertial force—each piece of the cable wants to go in a straight line but is being pulled down by the cable. Its inertia is therefore pulling back up. As the cable gets longer from the ground up, the centrifugal force increases. At some point, the force of gravity and the centrifugal force balance. At this midpoint there is no tension on the cable.

Above the balance point, centrifugal force increasingly dominates. At the height of geostationary orbit, the centrifugal force pulling up is the same as the gravitational force pulling down. Anywhere above geostationary orbit, centrifugal force will be greater, pulling the cable upward. Therefore, a massive counterweight (as heavy as the cable) just above geostationary orbit can be used to hold the cable taught, or you could have a lighter counterweight farther above geostationary altitude.

Once in place, the system should be stable. The idea is that you could then have elevators that climb the cable from the ground to geosynchronous orbit. (For terminology, "geosynchronous" is any orbit with a period of one day, while "geostationary" also has to be a circular orbit directly above the equator.)

There are other choices to be made regarding the placement of a base station or stations. They could be at sea, either on a platform or even on a large vessel. The advantage here is that it would then be mobile and could even move the cable if necessary. Conversely, with a land base station, the potential advantage is that it can be built at high altitude, such as at the peak of a mountain. The benefit of altitude is that it reduces the total length of the cable itself, which reduces the needed thickness. A large tower could also be constructed, adding a compression design to the tensile one.

Can such an astounding feat of engineering even work? It all comes down to the strength of the cable. That limiting factor will ultimately determine if a space elevator is possible. Specifically, we are talking about tensile strength, the resistance to being pulled apart. We can calculate this several ways. The first is simply to use a uniform cross-sectional area, so the cable is the same thickness throughout.

As the cable gets longer, it also gets heavier, so we can figure out the maximum length of cable made from any particular material before it would break under its own weight. Really what we need to know is the specific tensile strength of the material, how much tensile strength per weight, which we can calculate by the density of the substance. The best materials would have a high tensile strength with low mass.

Let's take steel, for example. We can calculate the tensile strain on a steel cable 35,786 km long. The density of steel = 7,900 kg/m^3, which calculates out to a maximum tensile cable stress of 382 GPa (gigapascals). This is 242 times the ultimate tensile strength of steel. So, we would need a material that has 242 times the specific tensile strength of steel to work—something we do not currently have. Just making the cable thicker does not help, because that also makes it heavier. We need greater strength per mass (specific strength).

The other approach is to have a nonuniform cable thickness. The force on the cable is maximal at the geostationary anchor point and decreases the closer the cable gets to the ground (because it isn't holding up as much weight of cable). So, to minimize the total weight of the cable, we could make it thinner at the ground and then increase the thickness to whatever it needs to be until we get to geostationary orbit. Taking this approach, we can calculate how thick the cable would get at the geostationary platform, if we start with a 5 mm thickness at the ground.

For steel the thickness would be 10^{54} meters—which is bigger than the known universe. That is another way of saying that steel is simply too weak and will not work. For carbon fiber, the thickness would need to be 170 meters, and for Kevlar the thickness would be 81.3 meters. These are ideal numbers, not accounting for manufacturing flaws. These are at least possible, but not really feasible. Even if perfect, the cost of building such a massive cable would likely erase any benefit.

Clearly we need a new advanced material out of which to build our cable. As discussed in chapter 11 on materials, carbon nanotubes are crazy strong for their weight. If we run the numbers, then we find that a uniform cable of carbon nanotubes would tolerate a strain of 130 GPa. This is much better than steel, but not quite up to the 382 GPa a space elevator cable would need. If we again take the tapering approach, the needed thickness of the carbon nanotube cable at geostationary orbit would be 6.37 mm. That's it.

Problem solved, right? The cable could be very thin and still be strong enough to connect a geostationary space station, even with a counterweight, to the ground. This means that a space elevator is at least theoretically possible. The laws of physics allow for it, but the engineering challenges remain enormous.

First, we do not know how to manufacture carbon nanotubes at length. The record so far is about 1 meter. Manufacturing a cable many thousands of km long is simply not possible. Further, while the tensile strength of carbon nanotubes is fantastic, it is fragile in other ways. Even the slightest flaw could cause a nanotube to "unzip." This could potentially cause the nanotube cable to unwind and break apart. That would be bad. So, while technically possible, it remains to be seen if a carbon nanotube cable is ultimately feasible.

There are a couple of other candidates, also all allotropes of carbon. One is coiled single crystal graphene. This has a similar specific tensile strength to carbon nanotubes. Like nanotubes, though, we do not currently know how to manufacture it in bulk. Diamond nanothreads are also made of carbon but arranged in a tetrahedral shape, which confers great hardness as well as tensile strength. The same manufacturing hurdle remains.

In short, at present we do not have a candidate material that has the requisite specific tensile strength that we could manufacture in the necessary length at a worthwhile cost. Whether or not such a material will ever be discovered remains to be seen, and this is why, ultimately, we cannot predict when and if a space elevator will be feasible.

Assuming we can build a space elevator, how would people and supplies get up it? Using a traditional elevator design with lifting cables is not practical. Theoretical designs typically include crawlers that climb up the cable. These would put their own stress on the cable, but that could be balanced with multiple coordinated crawlers.

Climbing time would depend on the power of the crawler motors and the safety of their design. About four days to geostationary orbit is an average estimate. Due to weight, it would not be feasible to have them carry their own fuel. Using electrical motors recharged wirelessly or through lasers is one proposed solution.

A geostationary platform is an excellent jumping-off point for any destination to deep space—the moon, Mars, or elsewhere. Most of the energy expended for such destinations is used getting into orbit. But what if you are headed for a destination other than the geostationary platform, such as the International Space Station (ISS) or other low Earth orbit (LEO) stations? This gets complicated.

First, any jump off the cable would need to be higher than geostationary height; otherwise the ship wouldn't be moving fast enough to stay in orbit and would just fall back to Earth. A rocket launch would be needed to get up to the correct velocity and vector for insertion into the desired orbit. Doing so from LEO height on the space elevator would actually cost more than just taking a rocket from the ground, but you could potentially climb to a greater height and then rocket down to the desired orbit from there.

Therefore, the utility of a space elevator would not be as significant as it may at first seem. Even with a functioning space elevator, low Earth orbit may still belong to rocket technology.

Cost versus Benefit

Assuming we can solve the engineering obstacles, would it ever be worth it to build a space elevator? Let's start with our primary motivation: to reduce the cost of getting stuff into orbit. To put things into perspective, the cost of getting 1 kilogram of material into LEO for the Space Shuttle system was $60,000 (averaged over the life of the program). A 100 kg astronaut plus gear cost $6 million to reach the ISS.

Since the Space Shuttle program, however, there has been a concerted effort to reduce the cost of getting into orbit. NASA, for example, decided to partner with private corporations and essentially privatize LEO so that they could focus on deep-space missions. One of the reasons for this move was specifically to allow competition, efficiency, and scale to reduce cost. That plan has worked out pretty well, including new companies such as SpaceX, Blue Origin, and Virgin Galactic, and also established aerospace companies like Boeing and Northrop Grumman.

The specific cost to orbit varies based upon the rocket, with heavier rockets overall being more cost-effective (higher percentage of the weight being lifted into orbit is cargo). SpaceX's Falcon Heavy, as of 2020, had a dramatically decreased estimated cost of $1,500 per kilogram. These savings are largely achieved through reusability. There is every reason to conclude that these costs will continue to drop as competition increases and the technology improves. SpaceX hopes to get the cost to less than $1,000 per kilogram, but it's hard to say where they will level off. At some point, further improvement will lead to diminishing returns as we run into the ultimate limits of rocket technology.

Estimates for the price to get a kilogram of material into LEO using a space elevator varies from $200 to $500 per kilogram (based on an estimated cost to build of between $6 and $40 billion). There are obviously many variables here, including the cost of construction, maintenance, and the useful life span of the space elevator. Let's say the lower estimate is accurate, $200 per kilogram, and rocket technology levels off around $1,000 per kilogram. This would mean the price for a person (again let's assume 100 kg for the mass of the person plus their gear) would drop from $100,000 to LEO to $20,000. That is significant, bringing down the cost of space travel into the zone of space tourism.

However, at the other end of the uncertainty range, the cost advantage might mostly disappear. If it costs closer to $500 per kilogram for the space elevator, and the space rocket industry gets the cost substantially lower than $1,000 per kilogram, the cost advantage becomes small and may vanish completely. This will also depend on orbital destination. Even if the space elevator is cost-effective to geostationary orbit, and deep-space destinations from there, it might ultimately be more expensive to use it for LEO destinations.

Safety is also a concern. Rocketing into orbit is now routine but still significantly riskier than taking a commercial jet flight. As flights become more common, the safety will improve. How safe would a space elevator be, though? That depends partly on how strong the cable is. The calculations we just did are based on the static tension on the cable. The cable would also need to accommodate dynamic stress from crawlers going up and down them, from weather, from the Coriolis effect (a lateral force on any rotating system), and even from space debris and small meteors. The crawlers themselves would need to be highly reliable, while traveling for days at 200 mph up a giant cable.

Another safety concern is that a space elevator would make a tempting target for terrorism and would be extremely vulnerable to sabotage with potentially catastrophic effects. This was dramatized on the Apple TV adaptation of *Foundation*, in which terrorists explode a bomb on Trantor's space elevator, causing the cable to fall to the ground, carving a giant scar that wrapped around the planet.

Convenience is a final consideration. If sending supplies into orbit, a four-day travel time is no big deal. However, for passengers, riding in a crawler for days, rather than hours for a quick rocket trip, may not be worth the reduced cost (depending on how much that is).

When taking all of this into consideration, it seems that the space elevator is one of those future technologies that looks great on paper and is theoretically technically feasible, but at the end of the day may not be worth the trouble. Rocket technology may simply improve too quickly for a space elevator to ever be sufficiently cost-effective enough to be worth the risks.

Still, the day may come when a cost-effective material exists with sufficient specific tensile strength and flexibility, so that a space elevator becomes practical. This presumably is a day in the future when demand for space travel is significant. I don't think this day will come anytime soon, however.

Alternate Space Elevators

The discussion has focused entirely on what some now call a "classic" space elevator, from the equator on Earth to geostationary orbit. But the concept could apply to other situations. Some have proposed a space elevator on the moon, on Mars, and theoretically to other worlds in the solar system we may one day inhabit.

A moon elevator could not go to lunarstationary orbit because the moon is tidally locked to Earth and rotates too slowly (once a month). However, the orbital anchor could be at a Lagrangian point, which is one of several gravitational low points when two or more gravitational fields overlap. Any direction from that point is essentially going uphill, and so things tend to stay there. This would be an efficient way to get off the lunar surface. However, the moon's low gravity, one-sixth that of Earth, means rocketing off the surface is not difficult or expensive anyway.

Mars might be in the space elevator sweet spot. Surface gravity on Mars is 38 percent of Earth's, but the day is almost the same length, 24 hours and 37 minutes. The martian geostationary orbit altitude is only 13,634 km, so the cable would be much shorter than for an Earth space elevator (35,786 km). The cost-benefit to a future Mars settlement may therefore be worth it.

In a 2019 paper, Zephyr Penoyre and Emily Sandford proposed a cable from the surface of the moon down to Earth orbit. People and supplies would still rocket into Earth orbit, but then could climb the rest of the way to the moon up the lunar cable. The lunar elevator could also be used to get from the moon's surface into Earth orbit.

There are also a variety of space-tether ideas, using long cables for momentum transfer or hauling objects into different orbits. One interesting proposal is called a "rotovator." This involves a station in LEO with a cable attached that is rotating in the direction of its orbit. The cable therefore swings above the station and then below the station and is just long enough to reach the ground. If the timing is synchronized well enough, the end of the tether would be briefly stationary on the ground, giving enough time to attach a payload that would then get hauled up into orbit. This seems like a delicate dance with far too many things that can go wrong, however. Imagine a giant cable swinging from orbit down to the ground.

It's interesting to speculate about the possible utility of giant cables in space travel. It's a way for space to be at our fingertips. While the prospect of a classic Earth-based space elevator seems slim...it may be a viable technology in the far future and it's certainly fun to imagine.

FUTURE FICTION: 2511 CE

The UES *Spyglass* was approaching its destination, a completely empty patch of space 650 astronomical units (AU) from Sol. This was past the Kuiper belt, past the heliopause, in the vast nothingness of interstellar space. But it was still less than halfway to the beginning of the Oort cloud, so there weren't even cometary balls of ice to greet them.

The commander sat among an array of holographic displays, calculating deceleration rate and time to destination. He wished he had a porthole to peer through. It was quaint, but he liked the visceral feeling of looking directly out into space. But the bridge was by necessity at the center of the ship, shielded by a double layer of nanostructured foam metal. The engineer beside him had no such primitive longings, preferring a neural-link, allowing her to essentially become the ship.

The third member of their shift entered the bridge and said, "Our sleeping companions all check out. I foresee no problems, so we should be good to end the hibernation cycle in three hundred seventy-four hours. That will have them up and about by the time we settle into solar orbit."

The entire crew was genetically engineered not just for the rigors of space but also to facilitate years of hibernation. Three crew members were awake each shift, looking over the operation and their companions. Soon, though, they would be all awake for the first time, ready for their five-year mission.

Not getting a reaction, they poked a bit more. "I still say it was pointless to bring actual humans on this mission. The robots can handle everything just fine."

"Thank you," said Arturo76 from the back of the bridge. "Your confidence is appreciated."

That got a snort from the engineer. She shifted focus to her physical surroundings. "It's all political bullshit. Everyone is afraid the bots will form their own army and come back to take over the planet, so we're here to babysit. Hell, I have more AI in me than that thing," she quipped, gesturing toward Arturo76.

The *Spyglass*'s three fusion engines were currently burning at near capacity, creating 0.83 G of acceleration. It was a less-romantic version of flight than the initial acceleration away from Sol using their reflective, almost shimmering light sails propelled by powerful lasers. Out here past the reaches of human civilization, there were no lasers to slow them down again—that was one of the first things they would remedy—so they were burning hydrogen to decelerate to their new orbital velocity.

The commander finally turned to his crewmates. "Strap in. We were waiting for you to power down the engines."

Once everyone was in place, the engineer began the process of cycling down each of the fusion engines in turn. After about ten minutes, when the ship was no longer accelerating, everyone could feel the familiar sensation of weightlessness. They were not at their destination yet; this was the last full system check prior to the last burn into their final orbit. With the ship now coasting, a swarm of surveillance bots detached from the hull and began the tedious process of making a detailed inspection of the exterior of the ship from various angles and distances.

The entire vessel was a massive beast several kilometers long. The crew compartment was actually the smallest component. The main construct was a telescope, waiting to be positioned at the perfect distance from Sol to use the star's gravity as a lens. This would allow unprecedented magnification, enough to get relatively high-resolution images of nearby exoplanets (with a little help from a megaqubit quantum computer), out to about 100 light-years. Data from this mission would prioritize future interstellar explorations for years to come.

The next biggest component of the ship was the future deep-space station that would accompany the telescope. Designed to be entirely self-sufficient (except for deliveries of hydrogen fuel every decade or so) and large enough to support thousands of scientists, it was an O'Neill cylinder shape. Now motionless and dark, it would spin up once it was separated and in position.

Finally, there was the high-powered laser to facilitate future travel to and from the *Spyglass* complex. It looked small compared to the entire ship, but in truth it was a gigantic machine.

In the back of the ship were the three fusion engines and their many tanks of compressed hydrogen, used as both fuel and propellant. Once in position, this part of the ship would be disassembled. One engine would be attached to the laser, while the other two would power the station. The telescope itself required little power, which could be managed by periodically recharging its batteries from the laser or even the station. They were way too far out for solar power to be of any use.

The commander monitored the inspection of his ship, even though it was unnecessary. He took over manual control of one of the cameras and paused over the crew compartment. This was actually a self-contained ship and was what he and his crew would use to return to the inner system once their mission was complete and replacement crews had arrived.

Perhaps he would stay, however. Life on the station—on the farthest outpost of humanity—would be interesting. Well, he had five years to think about it.

PART FOUR

The Future of Space Travel

When trying to imagine the future of humanity, whether to make an accurate prediction or develop a future sci-fi world, one of the biggest factors to consider is the extent to which humans will spread out into space. Some visions of the future see our civilization spanning the solar system, even spreading out to multiple stars, with people zipping about in faster-than-light ships. At the other end of the spectrum, humanity might remain largely confined to Earth, with little more than some industry and small stations or settlements off world.

These two starkly different pictures of our future depend largely on how optimistic or pessimistic we are about space travel. Trekking through the astoundingly vast distances of space is an extreme technological challenge, and humans are simply not adapted to space. And yet we are already pushing out into space and sending our robots ahead of us.

It's hard to imagine the future without space travel, which has become almost synonymous with the future. This is why an image of a settlement on the moon or Mars is immediately recognizable as a depiction of the future and why so many science fiction stories either take place in space or heavily involve space travel. But is this a realistic or a romantic view of our likely future?

We're going to turn our skeptical eyes toward the future of space travel to see how plausible our hopes and visions for the final frontier are.

21. Nuclear-Thermal Propulsion and Other Advanced Rockets

The only way to get our asses to Mars.

The cornerstone of space travel is, of course, the spaceship. We need a way to "Slip the surly bonds of Earth" (as pilot and poet John Magee put it) and not only get into space but also travel through it to impossibly distant destinations.

There are two realities about space that must be confronted for any meaningful discussion of space travel to occur. The first is that it is an incredibly hostile environment for fragile humans. We cannot survive the near vacuum for more than about ninety seconds (and would likely pass out after fifteen). We need protection from the extreme cold (near absolute zero), or extreme heat if in direct sunlight. Outside the protective cocoon of Earth's atmosphere and gravitational field, we would also be bombarded with dangerous radiation. And there are no convenient places to pick up food or water in space, and no refueling stations. We would have to bring everything we need with us and build new space infrastructure as we go.

Second, space is really, really big. Mars, the nearest planet we have any hope of sending people to anytime soon, gets only as close as 62 million kilometers from Earth. It would take about 7 months to travel there with current rocket technology. The nearest star to our solar system is Proxima Centauri, about 4.2 light-years away, or more than 39 trillion kilometers. A vast distance that would take conventional rockets tens of thousands of years to cross.

Clearly interstellar travel will have to wait for advanced spaceship technology; meanwhile, where will our current rockets get us? Is the future depicted in the TV series *The Expanse*, based on the books by James S. A. Corey, which imagines a future three centuries from now, plausible? Using only rockets, can we eventually zip around the solar system, with settlements not only on the moon and Mars but also the asteroid belt and the moons of the outer planets?

The best way to predict the future of space travel is to compare the various technological options. They all have their own strengths and weaknesses, and no single option is perfect. We will likely have to combine a number of technologies to build a spacefaring infrastructure.

Conventional Rockets

The basic function of any rocket is what is known as a "reaction engine," because they follow Newton's third law of motion—for every action, there is an equal and opposite reaction. The idea is simple. You throw something (a propellant) out the back end of a rocket and the rocket is pushed in the opposite direction with equal total momentum.

Chemical rockets accomplish this by burning fuel (such as hydrogen) with oxygen. The burning chemicals turn into a rapidly expanding hot gas that escapes through the rocket nozzles. This produces thrust, pushing the rocket away from the direction of the escaping gas. For chemical rockets, therefore, the fuel is also the propellant.

We know this technology works: It's powerful enough to get large rockets into orbit and to send ships and probes to the moon and to all the other planets in the solar system. We also know that it is incredibly inefficient, for a few reasons.

The main limiting factor of this approach to space travel, specifically rockets that carry their own fuel and propellant, is known as the "Tsiolkovsky rocket equation." Essentially, the rocket has to carry enough fuel to lift the rocket and its payload to the desired destination. It also needs enough fuel to lift the fuel itself, and fuel to lift that fuel, and so on. Here is the equation itself: mass ratio $= 2.2^{(\text{delta-v/exhaust velocity})}$. Mass ratio is the entire mass of the rocket (rocket, cargo, and fuel) divided by the "dry" rocket mass without fuel. Delta-v is simply the total change in velocity of the rocket if all the fuel is burned. Exhaust velocity is the velocity of the propellant, which relates to how much velocity the rocket will gain with each amount of propellant/fuel used.

There are a couple of implications to this equation worth emphasizing. One is that the greater the exhaust velocity, the less fuel needed for any desired trip, such as getting a specific payload to the moon over a certain period of time. This is because momentum is equal to mass × velocity, so the greater the velocity of the propellant, the more momentum it imparts to the rocket at any given mass. This is also called "specific impulse"—the change in a rocket's momentum at a given mass of propellant. The only other way to increase thrust is to increase the mass of the propellant, but that is not optimal because it brings you back to the rocket equation—you need fuel to carry the extra mass.

The most efficient rocket would therefore use fuel that produces a propellant with an exhaust velocity near the speed of light (the fastest anything can go). The slower the exhaust velocity, the more mass you need for the same thrust. In short, the faster you can get propellant to fly out the back of a rocket the better, and the less of it you need.

Also, the amount of fuel needed for increasing dry rocket mass or desired delta-v, increases exponentially, not linearly. There are several dramatic ways to illustrate this. Let's say that Earth's surface gravity was 1.5 g instead of 1 g. In order to get into LEO, you would need a greater delta-v, also called "escape velocity." If you do the calculation, no chemical rocket could produce enough thrust to get a rocket, even without cargo, into orbit. A civilization on a planet with 1.5 or greater surface gravity would never be able to get into space with chemical rockets.

As another example, right now, we have probes like Pioneer 10 and Voyagers 1 and 2 heading into interstellar space. At their velocities (which resulted from a combination of rocket thrust and gravity assists, so not even pure rockets), they would take tens of thousands of years to get to the nearest stars (if they were headed in that direction, which they are not).

Let's say we wanted to get there more quickly. What if we wanted to send a single toothpick to Proxima Centauri 4.2 light-years away and have it arrive in 100 years? That would require a greater delta-v to make the trip that quickly. With chemical rockets, it would not even be possible—the trip would require more rocket fuel than is available in the known universe, more than the mass of 2 million Andromeda Galaxies.

Clearly, we won't be using chemical rockets for interstellar travel. They can be useful for travel within our own solar system, but even there, they are extremely limited. Seven months to Mars means you won't be going there for a two-week vacation, and any settlement on Mars would be more isolated than the New World colonies were from their European brethren. There is definitely room for improvement, but there are still theoretical limits of chemical rockets and other options to consider.

For any type of rocket engine, we must look at the specific impulse. Technically, specific impulse is the change in momentum produced from burning an amount of mass (pounds or kilograms) over one second and is measured in seconds. The reason for the units is that specific impulse equals the number of seconds that the initial mass of fuel would be able to accelerate itself at 1 g. The higher the specific impulse, therefore, the greater the efficiency of the engine. For chemical rockets where the fuel is the propellant, specific impulse is directly proportional to the nozzle velocity of the propellant. This also necessitates knowing the maximal thrust.

Some engines, for example, like ion engines, have a very high specific impulse but can produce only very tiny thrust. They are efficient but it would take a very long time to get up to any significant velocity (even if you had twin ion engines).

Another factor when considering the theoretical limits of different engine types is the efficiency of conversion of fuel to energy for thrust. As we discussed in chapter 16 on energy, different types of reactions have different efficiencies in terms of energy production. For this comparison, we will use megajoules (MJ) of energy produced per kilogram (kg) of fuel. (A megajoule is 1 million joules, with a joule being 1 watt of energy for 1 second.) Chemical fuel has a potential energy production of 1–5 MJ/kg. Solid rocket fuel is more energy dense, at 5 MJ/kg, while liquid fuels tend to be around 1 MJ/kg. Nuclear fission has 8×10^7 MJ/kg, while fusion has 3.5×10^8 MJ/kg. Finally, antimatter fuel could potentially release 9×10^{10} MJ/kg—10 billion times the energy per mass as chemical reactions.

The most efficient chemical fuel/propellant is hydrogen oxygen. Even using that, to get a rocket into low Earth orbit, 83 percent of the mass must be fuel. The Saturn V rockets that took astronauts to the moon were 85 percent fuel on the launchpad. We are therefore close to the limit of chemical rockets. If we want to do significantly better, we will need to make the change to nonchemical propulsion.

Advanced Engines

Conventional chemical rockets burn fuel, causing the fuel to rapidly expand and get expelled as propellant. Chemical rockets generally have a specific impulse of about 450 seconds. The advantage of chemical rockets is that they have high thrust—the Saturn V rocket produced 35 million newtons of force (1 newton accelerates 1 kg mass 1 meter per second). However, we can separate the energy source from the propellant, allowing for the use of fuel with more energy per mass.

Solar thermal rocket designs use concentrated sunlight to heat a chamber of hydrogen, which then heats, expands, and is expelled as propellant. Laser thermal rockets are similar but use a high-powered laser instead of sunlight to heat the hydrogen. The advantage here is that the energy source is external and unlimited. Such engines would also have a higher specific impulse than chemical engines, greater than 800 seconds. However, they have modest thrust, about 3 newtons. They would therefore be useful for flying between planets, but not getting into orbit in the first place.

Thermoelectric engines use electrical resistance to heat propellant. They could therefore work with any source of electricity. They also have high specific impulse, 800 seconds, but low thrust, about 1 newton.

The aforementioned ion engines use either electrostatic or magnetic fields to accelerate ions, which are charged particles. The ions are accelerated to very high velocities, and therefore such engines have a high specific impulse, up to 10,000 seconds (20 times a chemical rocket). However, we can't forget that low thrust, about 3 newtons.

Some of these types of engines are already in use. They may not be very capable of or used for lifting out of Earth's gravity—we still need the high thrust of chemical rockets for that—but they are highly efficient for moving things like probes around the solar system.

We don't currently have a nuclear thermal propulsion (NTP) engine, but we do have possible designs for one. In fact, NTP technology was being developed by NASA in the 1960s, but the program was ended due to funding. The big advantage of NTP is that uranium has 4 million times the energy density as typical chemical rocket fuel. A fission reaction can be used to generate heat, which will then heat hydrogen as a propellant (up to about 2,500°C). The intense heat generated by a nuclear reactor could result in double the exhaust velocity, and therefore roughly double the specific impulse to around 900 seconds.

How much thrust an NTP system would produce is not clear at this point, but it would be less than what is needed to get into low Earth orbit. So again, an NTP system would be a combination of chemical rockets for launch, and then NTP would produce the necessary thrust to get to Mars or other deep-space destinations. Because of the high efficiency, such systems would be able to carry a relatively greater percentage of cargo. Further, the total travel time would be decreased. Estimates are that using an NTP system travel time to Mars could be cut in half, from six to nine months down to three to four months.

The hope is that with improved technology, NTP systems could become still more efficient and powerful. While we are getting to the limit of chemical propulsion, once developed we would be just at the beginning of the potential of NTP technology.

Nuclear electric propulsion (NEP) systems use nuclear fission to drive a power plant to generate electricity. The electricity is then used to accelerate ions in the same way as the ion engines described above. So, NEP is simply a nuclear-powered version of an ion engine. NEP systems would also be high efficiency with specific impulse up to 10,000 seconds, but low thrust. They could be useful, however, for cargo missions or probes to distant locations.

In the more distant future, there is also the possibility of fusion engines. Like fission engines, they could use thermal or electrical propulsion systems. Fusion systems have the potential to produce twenty-five times the energy per fuel mass as fission-based engines. They also have the advantage of not producing radiation that any hypothetical crew would need to be shielded from. Otherwise, the parameters would likely be similar, with sustained fusion simply being another way to generate heat.

There is another theoretical way to produce thrust with nuclear fission or fusion other than a sustained controlled reaction. While this seems rather crude, we could explode either fission or fusion bombs out the back of a rocket, which would be fixed with a very tough plate connected to a powerful suspension system. When the nuclear bombs explode, the force of the explosion will push on the plate and therefore transfer some of their energy to the rocket, increasing its momentum. This system is called a "nuclear pulse engine." Different theoretical versions of an NPE could produce a specific impulse of between 10,000 and 100,000 seconds and could reduce travel time to Mars down to 4 weeks.

One interesting design for a futuristic spaceship engine is the direct fusion drive (DFD). This system uses a unique magnetic confinement and heating system to fuse helium-3 and deuterium. It can generate electricity, which can then be used to power the ship's systems. The system can also directly produce thrust by adding propellant to the edge of the plasma flow in the fusion reactor. Propellant will be exposed to tremendous heat, becoming ionized and then accelerated through a magnetic nozzle. This would essentially function as a form of ion thruster and could, therefore, produce a high specific impulse. A theoretical DFD system could propel a 1,000 kg cargo to Pluto in 4 years.

The theoretical limit of fuel efficiency, of course, would be realized with matter-antimatter engines. Matter and antimatter annihilate each other with 100 percent conversion of mass into energy—you can't get any better than 100 percent. There is a lot of energy in mass (remember $E = mc^2$). However, not all of this energy is easily utilized, as about half escapes as gamma rays and neutrinos. Combining 1 gram of matter with 1 gram of antimatter would equal 1.8×10^{14} joules of energy, or about 43 kilotons.

There are two obvious ways to use this energy for propulsion. As with nuclear engines, you can simply use the heat produced to heat a propellant like hydrogen. This is actually a fairly efficient use of the energy. Or you could engage a strong magnetic field to accelerate the ionized products of the annihilation, producing a powerful ion thruster. Antimatter engines are not yet feasible, mostly because we do not know how to make significant amounts of antimatter.

For now we are stuck with chemical rockets and the limits placed on them by the rocket equation, but (if we choose to develop them) we can be on the cusp of nuclear engines. Fusion engines are plausible probably for next century, while matter-antimatter engines remain in the very distant future. Therefore, for a very long time, our future in space with be dominated by chemical and/or nuclear rockets. What does this mean for our vision of the future?

It means that space travel is going to suck long into our future. We can send our probes all over the system, but people will likely be confined to the Earth-moon system. A Mars settlement is plausible but will remain extremely isolated from Earth, with several months' travel time being the best we can hope for. From this perspective, our future in space is likely to be far more pessimistic than most sci-fi or future visions.

But is there any way to rescue us from the rocket equation, and to have something at least a little closer to what science fiction and futurists have been promising for decades? Let's consider propulsion systems that do not carry their own fuel or propellant.

22. Solar Sails and Laser Propulsion

Why carry fuel if you don't have to?

In a 2012 article titled "The Tyranny of the Rocket Equation," NASA flight engineer Don Petit wrote: "If we want to expand into the solar system, this tyranny must somehow be deposed." The fact is, rocketing around the solar system is always going to be a frustratingly slow process. Even with advanced nuclear rocketry, getting to Mars would take weeks and going farther to the outer planets would take years. Traveling to even the nearest stars would not be feasible.

In the aforementioned excellent TV series *The Expanse*, in order to explain how humanity has spread throughout the solar system, they had to invent a mysterious rocket engine that would allow for amazing thrust and specific impulse, as well as a drug to allow people to tolerate crazy g-forces. It's a common sci-fi plot device, a "gimme" to facilitate the story, but it's hard to imagine even theoretically what such a rocket would be. *The Expanse* is otherwise a very hard science fiction look at the future, so it's telling that they had to invent a fictional rocket design to make it plausible.

To truly get around the rocket equation, we need a ship that does not have to carry its fuel and propellant around with it. This leaves us with two options: either the ship needs to be pushed from the outside, or it needs to scoop up its fuel/propellant as it goes.

Solar Sails

The concept of a solar sail is quite simple, as it functions much like a regular sail on a ship. Instead of being pushed by wind, however, a ship with a solar sail is pushed by light from the sun. Despite having no mass, photons of light do have a tiny amount of momentum. If a photon is absorbed by a material, then that momentum is transferred to the material. Better yet, if the photon bounces off the material, then twice its momentum is transferred—enough to not only stop the photon but also send it back in the opposite direction at the speed of light. Reflection also has an advantage over absorption in that it reduces heating of the solar sail material.

The idea of a solar sail was first formally proposed in 1966 by György Marx in a paper published in the journal *Nature*. The concept was first tested by NASA in 1974 with their Mariner 10 probe to Venus and Mercury. Once the probe was out of fuel, they turned its solar panels toward the sun and were able to detect a small propulsion from the sunlight, serving as a proof of concept.

It wasn't until 2010, however, that an actual constructed solar sail was deployed in space, by the Japanese Space Exploration Agency's Interplanetary Kite-craft Accelerated by Radiation Of the Sun (IKAROS) spacecraft. The craft deployed a 14-meter sail and demonstrated it was able to control its direction. It also measured a tiny thrust of 0.2 grams, but that was enough to move the craft over time.

In the same year, NASA tested a prototype solar sail of their own, called NanoSail-D. This was only a 3-meter sail deployed on a small satellite in Earth orbit. The craft burned up in the atmosphere after 8 months. The same fate befell LightSail-1 launched by the Planetary Society in 2015. This had a Mylar sail of 32 square meters (a size often compared to a boxing ring). The sail was used to extend the orbit of a small satellite, which it did, but still met a fiery death in Earth's atmosphere.

In 2019, however, the Planetary Society launched LightSail-2, which contains a similar sail and a small CubeSat. The craft is still in low Earth orbit, but higher than the orbit of the ISS. At this height there is still too much atmospheric pressure to keep the orbit of the satellite from slowly decaying, but the solar sail is extending its orbit by orienting toward the sun twice each Earth go-round and giving a little push. The primary purpose of this mission is still to learn about the basic technology of light sails.

Craft propelled by light pushing against a large reflective sail may seem low tech, but it is actually the fastest type of craft we currently have the technology to build, or even design. It is the only type of craft with any hope of getting to another stellar system in a human lifetime. The advantage of not having to carry around fuel or propellant simply leaves all previously discussed rocket designs in the dust.

Even better than relying on sunlight for propulsion, light sails could be pushed along by powerful lasers. This would work the same way as solar sails and has the advantage of lasers being more powerful than sunlight as well as being whatever wavelength of light is optimal and directed in any direction we wish. To be clear, solar sails can tack like a regular sailing ship, so they can theoretically go in any direction, but tacking comes at the cost of efficiency.

Laser-propelled light-sail ships could get to the outer planets in mere months compared to years for even the best rocket and could accelerate to 10 to 20 percent the speed of light (depending on their size). Maximum velocity is largely a matter of drag as the sails push against the interstellar medium. At some point drag equals the push from the laser and the craft will not go any faster. This is still very fast. In fact, there are already designs (the Breakthrough Starshot initiative) for sending small microchip probes to Alpha Centauri using laser-propelled light sails, with a travel time of twenty years.

Solar/light sail technology still has a long way to go, but it can work even today in a basic form. One current limiting factor is the sail material. An ideal light sail would be very thin, durable, lightweight, and highly reflective in a range of wavelengths. It would have to be durable enough to withstand plowing through the dust of space at high speeds, without creating long tears. LightSail-2, for example, has rip blockers in the Mylar fabric, so that any small tear will stop and not continue along the sail.

When using powerful lasers, it is also critical that the material can reflect enough of the laser light so as to not heat up significantly. We don't currently have a perfect material, but there are some candidates that might work, such as crystalline silicon and molybdenum disulfide. There are also suggestions to coat a light sail with sapphires to provide sufficient reflection.

Another limitation are the lasers themselves. In 2019, the Thales team of French and Romanian engineers demonstrated a 10 petawatt (PW) laser that operated for four hours. This is the level of power that would be needed for light sail craft with significant cargo or human passengers. Larger vessels would require even more powerful lasers, or a cadre of lasers.

Powerful enough lasers could also be used to partially clear the flight path of debris. They could provide energy for the ship itself, further reducing the need to carry fuel and therefore reduce the total weight of the ship.

However, as the pushed craft accelerate faster and faster away from the laser, the light will become increasingly redshifted due to the Doppler effect. This means the light sail would have to function well for increasingly longer wavelengths of light. Accurately targeting a craft at increasing distance will also be a challenge (like hitting a bull's-eye at millions of miles). Finally, no laser is perfectly coherent, so the beam will spread out increasingly over distance. These factors will all place practical limits on laser light sail propulsion.

Another practical limitation arises when sending a craft to a new solar system, as there are no lasers in that system to slow the craft down. We would need to match the vector of the destination star system, but since we want to get any interstellar craft up to relativistic speeds, they would almost certainly have to significantly slow their velocity in the direction of the destination star.

Science fiction author and physicist Robert Forward has

proposed including a detachable mirror in a light-sail ship. When you need the ship to slow down, or accelerate back home, you detach a mirror in front of the craft that reflects a laser beam back at the sail, slowing down the ship or accelerating it back to Earth.

Light-sail ships could also accelerate to their maximum speed, then fold up their sail to reduce drag and coast to their destination. Once they do need to slow down, they could deploy their sail to produce drag. If the terminus is another stellar system, then light from the destination star could also be used to slow the craft. Theoretically, a magnetic light sail could produce more drag for significant deceleration.

Once there, lasers could be installed to better slow further ships headed to that location. A laser sail–based system of space travel would get easier the more infrastructure we are able to put in place. We could, for example, have multiple lasers placed along any travel route. Such an infrastructure would be much easier within our own solar system, as rockets could be used to get lasers in position.

The basic technology of laser-based light sails seems doable without any major breakthroughs; therefore, given the tremendous advantage of light sails, it seems very likely that, rather than rocketing around the solar system in the next few hundred years, we will be sailing around the solar system. Also, light sails are by far the best prospective option for sending probes to interstellar space, including to nearby stars.

It's interesting that laser sails (or laser thermal propulsion, where at least the energy source is external) are so rarely depicted in science fiction and are often treated as quaint or a lesser option. In reality these are our best options for future space travel. The advantages of external energy and propellant are simply too great.

Ramjets

In 1960, Robert W. Bussard proposed the Bussard ramjet. This is a fusion rocket that scoops up hydrogen from deep space to fuel its fusion engine. The interstellar medium is 99 percent gas, and 75 percent of that gas is hydrogen while 25 percent is helium with an average density of about 1 atom per cubic centimeter.

The hydrogen would be fused into helium, producing tremendous energy and heat. Scooped hydrogen could also be heated and used as propellant. The scoop itself would have to be enormous, but it would be aided by a powerful magnetic field, kilometers wide, that will draw in hydrogen ions.

This large size would cause drag, just as with light sails. Therefore, the faster a ramjet goes, the more gas it can scoop up but also the more drag it experiences. Eventually these forces would balance, and the maximum velocity of the ramjet relative to the interstellar medium will be reached. This has been calculated at about 0.12 c (12 percent the speed of light). While still slow by interstellar standards, it would mean we could send such a ship to Alpha Centauri in about thirty-five years.

There are lots of technical details about the exact fusion cycle used, the exhaust velocity of propellant, and the density of the interstellar medium that ultimately affect the feasibility and potential velocity of theoretical ramjets. So, these numbers are a bit speculative. But if they end up being close to reality, this could be a viable technology.

Unfortunately, the most recent calculations are very pessimistic. In 2021, physicists Peter Schattschneider and Albert Jackson calculated that such a ship would require a magnetic field 150 million kilometers across and 1 AU deep (AU = astronomical unit, the average distance from the sun to Earth). The stresses on the ship itself would be immense, limiting the ultimate speed of such a ship to perhaps about 20 percent of the speed of light. In their paper they conclude that "it is very unlikely that even Kardashev Civilizations of type II might build magnetic ramjets with axial solenoids." Yikes.

There are also other theoretical possibilities for external fuel designs. For example, if dark matter turns out to be a particle that, like light, is its own antiparticle, then it might be a candidate material for scoopable fuel. Of course, we can't make any predictions about this until the nature of dark matter is discovered.

Black Hole Drives

It is theoretically possible to power an interstellar ship with a black hole. First, you need to create an artificial black hole of the right size. (Stay with me here; this is not as outrageous as it sounds.) This can be done without discovering any new physics—it just takes a tremendous amount of energy. You can use multiple converging megalasers, or high-energy collisions or implosions. Such a created small black hole is called a "kugelblitz black hole." The technological difficulty here would not be trivial, of course. This is something for the far future, but physics allows for it.

Also, no one is proposing this as an energy source on Earth. It's just too dangerous. In addition to being, well, a freaking black hole, a kugelblitz black hole is extremely hot. So hot that we currently do not even have the physics to describe how something that hot will behave. It's probably a good idea to keep them far away from Earth, on interstellar spacecraft. But a deep-space kugelblitz black hole could theoretically be used to power a solar system–wide civilization.

Anyway, once you have even a tiny singularity, then all you have to do is feed it matter. Any matter will do; we could use it as a giant garbage disposal. Physicist Stephen Hawking was the first to figure out that black holes give off radiation, appropriately called "Hawking radiation." At their event horizon, the point beyond which nothing, not even light, can escape, a virtual particle-antiparticle pair might be created at just the right distance that one particle gets sucked into the black hole while the other just barely escapes. You can't get something from nothing, so this particle carries away a little mass from the black hole.

The smaller the black hole, the more Hawking radiation it gives off and the faster the black hole loses mass. Eventually, at some minimal mass, the black hole just explodes in a shower of Hawking radiation.

Let's say we have created a tiny black hole, something around the size of a proton, which would have a mass of 606,000 metric tons. This would give off about 160 petawatts of power, or about 10,000 times the amount of energy the world uses in a year. Such a black hole would live for about 3.5 years before evaporating away. However, it could be fed more matter to keep it topped off, in which case it could survive indefinitely.

If we could safely pull this off, that would be a tremendous energy source. How could we harness that power to drive a spaceship? That gets thorny, and it depends on the trade-offs you want to make. One method, proposed by Lois Crane and Shawn Westmoreland of Kansas State University, is to surround the black hole with a parabolic reflective mirror that can direct the Hawking radiation out one side, to produce thrust for the ship. A powerful particle beam can be shot into the black hole to feed it matter to maintain its mass and to accelerate it along with the ship.

You could also surround the black hole with a substance to capture the radiation and turn it into enough heat to drive a heat engine. But then, of course, you need propellant, and we find ourselves back to the rocket equation.

There are massive engineering and logistical hurdles to harnessing an artificially created black hole, but it could produce long-term thrust, enough to accelerate a large ship to relativistic speeds. However, this is still definitely a theoretical drive for the distant future.

The Future of Space Travel

For now, we are dependent on rockets to get off Earth and reach destinations within our solar system. Nuclear fission and eventually fusion-based rockets are likely coming and will be a mainstay of space travel, perhaps for thousands of years to come. Light-sail technology will likely also be important and may be our only viable method of interstellar travel for thousands of years. Super-advanced ships that use matter-antimatter or black holes are likely relegated to the far future.

There are challenges to space travel beyond adequate propulsion. The biggest is radiation. Outside the protective atmosphere and magnetic field of Earth, there are two types of radiation that can damage astronauts. The first is solar radiation, comprised of charged particles. This type of radiation can feasibly be shielded with a thick hull or even just water or both for doubled effectiveness. The real risk from this type of radiation is its variability, which can come in storms. Astronauts would need sufficient warning to get to more highly shielded portions of the ship before they were hit.

The second type of radiation is much more difficult to deal with: galactic cosmic rays. As the name implies, this radiation is everywhere in the galaxy, even distant from stars. While cosmic rays are protons or electrons traveling near the speed of light, these are high-energy particles that are difficult to shield against. They have so much energy that they can even penetrate lead. Not only does existing shielding not work but it may also make the situation worse, trapping cosmic rays inside a ship causing more damage, or kicking off daughter particles and creating more radiation. Right now, we do not have a good solution to protect astronauts long-term from the steady rain of cosmic rays, which are powerful enough to damage DNA and cells. We are exploring biological treatments to reduce or treat the effects, but NASA's answer for now is to keep space missions as short as possible—essentially, "get there fast."

Another extreme challenge is how to create artificial gravity. This can be done with constant acceleration, but for the foreseeable future, spaceships will spend a lot of time coasting. This makes rotation the only feasible mechanism for generating artificial gravity.

Spaceships are generally too small for this. If the rotation structure is not huge, on the order of 1 km or more, then the rotation will generate vertigo. It is also challenging to rig equipment to work in microgravity and rotating gravity. Ships that are large enough and advanced enough to use rotation for artificial gravity are likely a long way away.

Again, if we are being realistic, and no matter how much it pains this Apollo and *Star Trek* fan to hear this, for a really long time space travel is just going to suck. Travel times will be long, there will be no artificial gravity, and travelers will likely be stuck in small massively shielded parts of the ship most of the time. We may, however, be partially saved by advanced materials, like metamaterials that create highly effective shielding from cosmic rays. Artificial gravity will come from acceleration only, which is why powerful fusion engines will be key for intrasolar travel.

Spaceships alone won't do it. We will likely make space travel more feasible and comfortable by building infrastructure. In the future, there will likely be a division of labor among various types of ships. We will likely develop large ships that continuously travel in an orbit from the vicinity of Earth to Mars or other deep-space destinations and back again. This ship could be optimized for comfort and with plenty of living/working space and supplies. A large enough ship can be well shielded and even have artificial gravity from rotation. These ships can be large and complex because they will only have to be accelerated once, and then will just coast on their circuit. They will be more like small cities in space, more space station than spaceship.

When traveling to Mars from Earth, a passenger could take a very small ship to rendezvous with the larger one, then dock so that they can spend the rest of the journey in greater comfort. Once at Mars, the smaller ship could then detach, inserting itself into Mars orbit or landing, while the larger ship continues on its trajectory back to Earth.

This infrastructure might also include a system of lasers to accelerate light sails around the solar system and to send solar ships on their journey to the nearby stars. But in the far future, we may harness the power of dark matter, black holes, and other exotic physics, getting us to 50 percent the speed of light or even higher. This would finally make interstellar travel truly feasible, even though trips to even the nearest stars will still take years to decades. However, traveling at greater than light speeds as in *Star Trek* and so much science fiction remains in that realm of imagination.

The infrastructure of space travel will also have to include more than just ships. We will need places to go, with resources and safe places to live, grow food, and stock water. We will not just explore and travel to space; we will have to settle space as well.

23. Space Settlements

Getting to Moon Base Alpha and Beyond.

Because humans evolved on Earth, we are highly adapted to this environment—a thin shell surrounding our planet with a narrow range of temperature and pressure, protected from radiation, and with an atmosphere with sufficient oxygen without too much carbon dioxide or other toxic gases, and the right amount of humidity. We evolved to see in the light put out by our sun and to eat the organic matter of which we are a part.

Outside this tiny protective cocoon, we would not survive for long. Even within it, there are places that we cannot survive indefinitely. The deserts are too hot and dry, the poles are too cold, the highest mountains have air that is too thin. So, we generally stay in our comfort zone, using clothing and even environment-controlled homes to remain truly comfortable.

The rest of the known universe, by comparison, is a deadly hellscape. In fact, there is no other place that we currently know of where a human could survive without a protective space suit or enclosed habitat. Despite the common science fiction trope that space travelers can land on most planets and be relatively comfortable, it is likely that only a very tiny percentage of even "earthlike" planets would be truly hospitable for us. There are just too many variables.

Other worlds would likely have too much or too little gravity, their sun would be too bright or too dim, the temperature would be too hot or too cold, and the atmosphere could have any number of things wrong with it. Without a magnetic field, we would be bombarded with deadly radiation. To top it all off, there is also likely nothing we could safely eat there.

Space itself is even worse. At temperatures near absolute zero, we would freeze but if we got direct, unfiltered sunlight we would boil. As previously mentioned, any even brief exposure to the vacuum of space would be lethal. Without shielding, the radiation would slowly kill us as well. Plus, we'd need to bring all our oxygen, food, water, and energy with us. Essentially in order to live in space, we would need to create a bubble that closely mimics conditions on Earth.

On top of all that, there is the lack of gravity. In orbit, which is a constant state of falling, there is essentially no felt force of gravity, what is technically called "microgravity" because there are still local sources of tiny gravity. Without something close to 1 g of gravity, our bodies do not function. We lose bone and muscle mass and can't distribute fluids properly, our vision is impaired, and we are still learning about other adverse effects.

Settling space is not going to be easy. We're still going to do it, because exploring and settling new locations is the human way.

Space Stations

It's a stretch to say that any existing space station is a "settlement" or even permanent presence. A settlement implies a self-sustaining community. It does not have to be completely self-sufficient—it can be dependent on outside trade or support—but should be a place where people live and work.

As an aside, space enthusiasts have stopped using the phrase "colonizing space" because the definition of a colony is when a foreign power establishes partial or full control of part of another country, with settlers from their own country. If there were actual martians on Mars, then an Earth settlement there would be a colony. Until we discover alien life, therefore, we'll stick with the now conventional "settlement."

Existing space stations, places where people can live and work in space, are not full settlements, but they are outposts or, as the name implies, stations. The ISS is currently the longest-lived occupied structure in space. It is also the largest at 108 meters long end to end, with as much interior living space as a six-bedroom house. Its main construction was completed between 1998 and 2011, although upgrades and repairs have been almost continuous.

The ISS has been continuously occupied by humans since November 2, 2000, including 242 different people from 18 nations (as of November 2020). American astronaut Scott Kelly holds the record for the longest continuous presence on the ISS , at 340 days. However, astronaut Peggy Whitson has spent the most time on the ISS—665 days total (but not continuous).

There have been previous space stations, including the Almaz and Salyut series, Skylab, Mir and Tiangong-1, that are no longer in operation. Together these stations have allowed us to develop the science and technology of living in space. Astronauts have even grown food in space, although not nearly enough to feed the occupants. The ISS is completely dependent on shipments of food and water.

Waste is only partially recycled on the ISS. Human feces is treated to kill bacteria, then eventually shot out of the station where it will burn up in Earth's atmosphere as a "poop meteor." However, NASA is looking for ways to fully recycle human waste. One proposal is to store the waste in an outer layer of a station to use as a "poop shield" against radiation. This idea was humorously explored in the comedy science fiction series *Avenue 5*, but it is a serious proposal. Urine has been recycled aboard the ISS since 2009, and the system has been upgraded with more efficient models that use strong acids to purify astronaut urine. As one astronaut put it, "Today's coffee is tomorrow's coffee."

The ISS is cleared to fly through the end of 2030. NASA plans to retire the station, dunking it into the ocean sometime in 2031, and replace it with a series of commercial stations. A private company, Axiom Space, is planning their own module to attach to the ISS with missions beginning in 2022. These modules are a substantial upgrade from the aging ISS, and once the ISS is decommissioned, the plan is for the Axiom modules to detach and become their own station.

Another private company, Orbital Assembly Corporation (OAC), has revealed plans for its Voyager station, which will hold up to 400 guests. This is a doughnut (torus)-shaped station much like the one from the movie *2001: A Space Odyssey*, designed to rotate to create artificial gravity.

As part of the Artemis mission, NASA plans to build a lunar space station called the "Gateway" to facilitate missions to and from the surface of the moon. As we develop the space infrastructure from low Earth orbit to the moon (referred to a "cislunar space"), there will likely be many more space stations in the future.

But again, stations are places to visit or to complete missions, not to settle. There are currently no plans to build a space station settlement. A permanently settled space station will likely not be feasible until the cost of getting into space lowers considerably and the technology of self-sufficiency in space is further developed.

For a space station to serve as a permanent settlement, it would need to be largely self-sufficient because of the expense and inconvenience of getting stuff into space. Fully recycling all water and waste would be a near requirement. The station would also need to be safe for permanent inhabitants, which means sufficient radiation shielding.

The ability to grow sufficient food for all inhabitants is also a plus but having supplemental food items delivered is feasible. Hydroponic gardens, growing plants in water without soil using completely controlled environments, artificial light, and almost completely recycled water is already big business on Earth (estimated at $9.7 billion in 2020). Hydroponics is an efficient way to cultivate many vegetables, and test plants have already been grown on the ISS.

Artificial gravity is also a must. Fortunately, this can be achieved with simple rotation. This would require a fairly large station, however. The bigger the station, the less likely the rotation is to be detectable and cause motion sickness. This torus design, like the Voyager, was first proposed in 1929 by Herman Potočnik, writing as Hermann Noordung, and for that reason is sometimes called "Noordung's wheel." However, there are other options. You could also have two modules connected by a long cable and rotating around each other. Ships could dock at the center point of the cable with elevators that climb down to one of the modules.

Another imagined design is a large cylinder rotating along its long axis, like the station in the show *Babylon 5*. This idea was first proposed in 1956 by Darrell Romick, who imagined a cylinder that was 1 km long, 300 m in diameter, and could hold 20,000 people. Gerard K. O'Neill published the first technical analysis in 1974, and for this reason the design is referred to as an "O'Neill cylinder." According to his calculations, steel would be strong enough to support a cylinder with an 8 km diameter, but not larger. More advanced materials could allow for larger structures still.

An energy supply is also critical, but this can easily be accomplished with solar panels. In space, solar panels can be oriented for a continuous optimal supply of sunlight. Battery backups would be necessary for emergencies or if the panels need to be taken offline for any reason. Other forms of energy in space are likely to be less convenient and cost-effective, and most likely used for backup power. But it is possible that a large future station might include a small nuclear reactor or even an advanced fusion reactor.

Oxygen is actually not as big a problem as you might imagine. If you are growing enough food to feed the inhabitants of a station, then those plants will also recycle the carbon dioxide and make more oxygen than is necessary. NASA calls this kind of system a "controlled ecological life support system." Perhaps the most famous experiment in such a system, although not optimized for space, is the Biosphere 2 closed habitat. A more relevant experiment was the Soviet BIOS-3, part of a research program from the 1960s to 1980s. They found that with 13 square meters per person, they could grow 78 percent of the food needed and nearly all the oxygen. With just a little more space or efficiency, this could easily create all food and more than enough oxygen.

In fact, simulations of possible Mars closed habitats find that excess oxygen could be a real problem. If the oxygen content of the atmosphere gets too high, then it becomes a fire hazard. Any closed habitat system would therefore need to carefully regulate the type and number of plants to perfectly balance food, water, oxygen, and carbon dioxide, or use an environmental system to tweak these variables. Likely it will be easiest to just overproduce oxygen then remove the excess, which can be stored for EVA (extravehicular activity) or ships visiting the station or used as fuel.

Other proposals for deep-space settlements include hollowing out large asteroids and then using the interior as a space station. If the asteroids are large enough, such as Ceres, then it may even be possible to "spin up" the asteroids (as was done on *The Expanse*—essentially using rockets to cause the asteroid to rotate more quickly) in order to produce artificial gravity. The asteroid itself would also provide ample shielding from radiation and meteors.

The bottom line is that a closed ecologically controlled habitat is not only perfectly feasible, but we are also pretty close to being able to accomplish this now. With such a system, sufficient starting raw materials, proper radiation shielding, some rotation for artificial gravity, and solar panels to power the whole thing, permanent space settlements are extremely feasible in the near future and likely to happen eventually.

This idea has a long history. The notion of a space station was proposed in 1895 by Konstantin Tsiolkovsky, considered the father of Russian rocketry. In 1903, he fleshed out his idea, including rotation for gravity, solar power, and a closed habitat for food and oxygen. A century and a quarter later, we now have the technology to at least begin to fulfill his prescient vision.

The real question is whether people will want to live in space. The answer likely is, given there are billions of people, someone will want to. But then we have to ask, what will people do in space, other than just live there? This is an economic question (and as we have seen, economic questions often trump purely technological ones). Could people earn a living in space, sufficient to afford what will likely be a high cost of living?

The economics of space settlements will depend on future industries. Even space stations with artificial gravity could have sections with low or even microgravity, useful for certain scientific experiments and perhaps critical for some future industries. Microgravity manufacturing may be the lifeblood of future space settlements. Mining asteroids, using a space station as a base of operations, is another possible industry. Asteroid miners will need a place to live and get supplies, as going back down and up the gravity well of Earth is not economical. A future space settlement may even house technicians whose job is maintaining a vast orbital solar panel array beaming energy down to Earth (or managing the robots who do the actual work).

Settling Other Worlds

We are not limited, of course, to building free-floating stations in space. We can use the same self-contained habitat technology to build settlements on the moon, Mars, and other worlds in our solar system. The issues remain largely the same. Such habitats need energy, food, oxygen, normal atmospheric pressure, and radiation shielding.

On a world like the moon, some of these things will be easier, some harder, and some the same. Food, oxygen, resource recycling, and atmospheric control are all almost identical technical problems in space or on the surface of a moon or planet. The only real difference is that it is easier to spread out on the surface of a world. You may also be able to use the local regolith to create soil for growing plants, although hydroponic farming is just as feasible. Some locations may even have a local water supply.

Solar panels are also likely to be an energy mainstay, as the continuous sunlight in space makes them extremely practical. The lunar poles can have continuous access to sunlight, but not the rest of the moon. This would necessitate building more solar panels and pairing them with sufficient battery backup systems.

Settling other planets is also where nuclear reactors are likely to be more valuable and necessary. They could more easily be built far away from living habitats and could produce continuous reliable energy for decades. Wind turbines are feasible on Mars, and in fact they would complement solar panels well, as dust storms that would block solar panels could create wind power.

The farther you get from the sun, the less energy is available for solar panels. For example, the maximum solar irradiance on Mars is about 590 W/m^2, a little more than half that of Earth at 1,000 W/m^2. This compares to only 50 W/m^2 at Jupiter. However, the outer solar system has lots of hydrogen in gas giants and on some of the larger moons as hydrocarbons. This could fuel fusion reactors or even hydrogen fuel cells (combined with that excess oxygen from all the food production). Even comets have volatile elements that could serve as fuel sources.

Radiation protection is easier on a solid world for two reasons. First, there are local building materials that can be used. Once you are on the lunar surface, you can construct your habitat, including thick radiation shielding, out of the local regolith.

From NASA images, we also now have good reason to conclude that there are natural caverns on the moon and Mars that could serve as perfect radiation shielding. These are lava tubes formed when lava came up to or near the surface. On the moon they are estimated to be from 300 to 900 meters wide (yes, wide) and many kilometers long, large enough to contain a small city. Because of the higher gravity, caverns on Mars are smaller but still range from 40 to 400 meters wide.

These deep caverns are protected from not only radiation but also from micrometeorites, or even larger meteors, which pose a real hazard to any space habitat without an atmosphere for protection. Further, we may be able to simply use inflatable habitats inside a lava tube, without the need for any heavy construction. It may even be possible to seal and pressurize the entire lava tube.

Perhaps the most difficult challenge for settlements on worlds with low gravity is that there is no easy way to increase the gravity to something closer to Earth normal. Ironically, in space with only microgravity, it is an easier problem to solve with rotation. On the surface of the moon, the gravity is only 0.165 g, and on Mars it is 0.38 g. At this point we do not know how we would even theoretically produce true artificial gravity, and it may not be possible. There are proposals to build large circular tracks so that a habitat can rotate, banked at an angle, to increase the g-forces, but this is likely to be an impractical solution.

We lack long-term studies of the biological effects of low gravity at lunar or Mars levels, but from what we do know, it is likely to reduce bone density and muscle strength among other negative effects. Perhaps it won't matter if you spend your entire life on the moon or Mars. You will become adapted to that gravity. However, it may make it difficult or even impossible to visit Earth. This may result in multiple human subpopulations, each adapted to different gravities.

Generation Ships

One other form of settlement worth mentioning is the generation ship. This is a spaceship large enough to serve as a space station, completely self-sufficient and able to contain hundreds or even thousands of people. The station, however, is also a ship, able to generate continuous thrust of about 1 g. In this case, the acceleration would provide the artificial gravity, rather than rotation. Alternatively, once the ship has reached its cruising speed, it can change configuration so that the living compartments are rotating to provide gravity.

Such a ship would be more challenging to design than a space station or moon base because it must be truly self-sufficient—there would never be any supply runs. This includes all maintenance and repairs. In deep space, there would also be no solar power, so fission or fusion power would likely be necessary, or something higher tech like antimatter or even a black hole engine.

Such ships are often called "generation ships" because they are designed for journeys that would take more than one human lifetime, decades to centuries. The people who arrived at the desired destination would be the descendants of those who left. Presumably the resources and people on the ship would then be used to start a new settlement in a distant stellar system.

Settlements in the Medium and Far Future

In the upcoming centuries, we will likely flesh out the infrastructure of our solar system, turning the moon into a giant city, fully settling Mars, building stations within asteroids, and eventually reaching out into the outer solar system. Large, well-shielded ships will crisscross the system connecting all these settlements together. By this time, our other technologies will have advanced as well. We may be using nanites and robots to autonomously convert asteroids into new worlds, making use of every scrap of matter in the system.

Earth will always be special, always our home, but we will eventually become a truly spacefaring species. That means not only building ships and settlements, but also changing ourselves to live in space. The people living in that fully spacefaring world may be augmented with machine components and artificial intelligence, and/or genetically engineered for the very different environments we will create around the solar system.

In 2022, researchers discovered that tardigrades (small water bears that can survive in harsh environments by drying out) have proteins that coat their DNA and protect it from radiation damage. Already these proteins or the genes that make them have been added to human cells in culture, where they make human DNA ten times more resistant to damage from radiation. Those genetically engineered people may also create new life to help them settle their new worlds.

If we keep going farther, spreading out to multiple solar systems, perhaps even ultimately the entire galaxy, what will this ultimately mean for human civilization? Another way to frame this question is: What might we find when we encounter advanced alien civilizations?

Once we can build and maintain habitats in space that are large and sophisticated enough, they may be more comfortable to live in than on any planet. This means that any advanced civilization can theoretically convert all the matter in the vicinity of their star into living space. Stars put out a tremendous amount of usable energy, so if a star were mostly surrounded by solar panels, almost all of that energy could be harvested to run a civilization. This approach was first proposed by Freeman Dyson and is therefore called a "Dyson sphere" (if the technology completely surrounds a star) or a "Dyson swarm" (if it partially surrounds a star).

In addition, the star could be surrounded by space habitats built out of the asteroids, moons, and even dwarf and full planets in the system. This would dramatically increase the living space available. A species confined to the surface of one planet may even seem quaint and primitive to such civilizations.

There are many reasons to move industry, food production, energy production, and even living space into outer space and onto or inside of otherwise barren worlds. The primary one would be to remove them from the biosphere of Earth, where they cause pollution and other damage. Eventually we may see the biosphere of Earth as a precious resource, best left mostly to natural ecosystems with a very light human footprint. Meanwhile, we can happily live off world, perhaps only visiting Earth's pristine nature.

It is difficult to predict what the long-term psychological outcome will be of living entirely in space. Will future generations find it natural, or will they pine to walk on the surface of Earth? Perhaps they won't care because they'll spend most of their time in a virtual reality, where they can walk on any surface they wish. While living in space or in habitat settlements may be fine, there will also likely be the allure of altering entire planets to suit our needs. This requires terraforming—engineering on a planetary scale.

24. Terraforming Other Worlds

Terraforming worlds may remain science fiction . . . for now.

In July 1942, writing under the pseudonym Will Stewart, science fiction author Jack Williamson published the novella *Collision Orbit* in the magazine *Astounding Science Fiction*. This story is credited as being the first to include the notion of terraforming and of coining that term. It has since become a staple of science fiction. In the *Alien* movies, for example, the Weyland-Yutani Corporation is involved in terraforming planets, including the one where the alien creature was encountered.

Sometimes "terraforming" does not go our way, as in the *War of the Worlds*. In the 1897 novel by H. G. Wells, the martians brought "red weed" to Earth, which rapidly spread on land. However, it was never clear if this was deliberate or accidental. In the 2005 *War of the Worlds* movie, it was made explicit that red weed was a deliberate attempt to "martianform" Earth.

As discussed in the previous chapter, the need for terraforming, or changing the characteristics of a planet so that they are more suitable for human life—more like Earth—makes sense. In 2013, using data from Kepler on exoplanets, the Center for Astrophysics—Harvard & Smithsonian estimated that there are likely 17 billion "earthlike" planets in our own Milky Way galaxy. Still, a tiny percentage of them are likely to have all the environmental conditions necessary so that a human could walk around with just normal clothing. There may be no such planets close enough to our solar system that we will be able to reach them even with advanced spaceships.

If we want to settle entire new worlds, with friendly environments and ecosystems, then terraforming is our only option. But is this even feasible? It depends on what the world is like to begin with.

Mars is the most likely candidate for terraforming in our own solar system. It's a little small, but we can live with that. In general, surface gravity will be the most difficult aspect of a planet to alter. If it is too high, there is pretty much nothing we can do in any reasonable time frame. If it is too low, we would need to add a lot of mass to make any significant difference. We could direct thousands of asteroids toward it, adding their mass to the diminutive planet, but that is a slow process and would likely make the planet uninhabitable for thousands or even millions of years.

Which brings up the most critical aspect of terraforming technology—it generally takes a very long time. This will be a limiting factor if we want to use the planet anytime soon, but not if we take a very long view.

So, what are the prospects for terraforming Mars? They are, to be blunt, not good. First, calculations show there is not enough volatile material on Mars to create a significant atmosphere. A 2018 NASA study found that even if all the frozen CO_2 and water vapor were released into the atmosphere, it would only increase the atmospheric pressure from about 1 percent (what it is today) to about 7 percent of Earth's atmosphere. That is barely enough to create enough pressure for a human to survive, and not enough for the greenhouse effect to warm Mars to a comfortable temperature. So even if Quaid did start the reactor (from the movie *Total Recall*), he and everyone else exposed on Mars would still die of suffocation.

Further, CO_2 is not enough. We need oxygen, which does not exist in significant amounts on Mars. We would also need a lot of nitrogen to get to a reasonable atmospheric pressure without too much oxygen or CO_2. However, we would not need to reach parity with one atmosphere of pressure, the pressure on Earth at sea level.

The summit of Mount Everest, for example, has about 0.33 of an atmosphere (depending on weather). This is barely survivable, and only by a few people with extreme conditioning and acclimation. Most climbers require supplemental oxygen. But it does show that it is within the range of what humans can withstand.

If a planet, however, had at least 0.5 atmospheres, and perhaps a higher O_2 content to compensate, that could be perfectly comfortable. Denver, Colorado, for example, has a pressure of 0.82 atmospheres.

What if oxygen were not an issue? What is the absolute lowest pressure that a human body can survive in before the blood will start to boil and your eyes will pop out like Arnold Schwarzenegger's character in *Total Recall*? We actually have a precise answer for this—0.0618 atmospheres, a pressure that occurs at about 63,000 feet above sea level on Earth. Technically it's the pressure at which water will boil at human body temperature. Pilot Harry George Armstrong was the first to recognize this phenomenon (don't worry, through physiological research, not personal experience).

Mars could get just above this pressure level if we melted all the CO_2 ice, but that still would not make the surface livable without aid. At that pressure, even breathing 100 percent O_2, you would not get enough oxygen to survive. For that you need 0.122 atmospheres with 100 percent O_2.

Another way to think about this is to consider the partial pressure of oxygen. Essentially you multiply the percentage of the atmosphere that is comprised of oxygen by the total pressure of the atmosphere. On Earth at sea level, the percent of oxygen is 21 percent and the pressure is 760 mm Hg, so the partial pressure of oxygen is about 160 mm Hg. Therefore, the lower the atmospheric pressure, the higher percentage of oxygen you need to have a normal partial pressure of O_2.

In addition to immediate survivability, there are also longer term adverse biological effects from low pressures, such as high-altitude pulmonary edema. This can be avoided with slow acclimation. Acclimation to low oxygen pressures can also happen over time, resulting in higher red blood cell counts, for example.

Ideally, a terraformed world would have enough CO_2 to keep it nice and warm, but not too much that it becomes unhealthy or fatal to humans. Earth's atmosphere has about 0.04 percent CO_2. At 0.1 percent, we can start to feel adverse effects, such as headaches and fatigue. At 5 percent CO_2, major health effects become life-threatening.

Air also requires humidity. If you have ever been to a desert, you'll have experienced how quickly one can get dehydrated. This is not only because of the heat, but also because of the very low ambient humidity. This requires water vapor in the atmosphere, which, to be sustainable, needs liquid water on the surface of the world.

So, at a minimum, a terraformed world needs an adequate atmosphere and a supply of water, with a temperature range that at least allows for liquid water on the surface some of the time. We may also need to remove toxic or corrosive substances from the atmosphere. Venus's atmosphere, for example, contains sulfuric acid, even small amounts of which would be deadly.

What methods are potentially available to achieve this end? For planets that have too little atmosphere, we would need to either release or deliver the constituents of a suitable atmosphere. Like Mars, some worlds may have frozen compounds like CO_2 that could be released by heating them up. This could be accomplished by simply melting them with heat generated by large and numerous power plants. Mirrors in orbit could also direct sunlight onto the ice to melt it. If possible, genetically engineered dark plants could propagate and absorb enough heat to melt the ice. Or we could redirect asteroids to crash into the planet, releasing tremendous heat.

There may also be compounds like oxygen and carbon dioxide bound up in the soil. This would be harder to release, requiring large processing plants, but possible. Mars, for instance, is red because of all the rust (iron oxide) in its regolith. If this oxygen could be practically extracted, it could feed the atmosphere.

Another needed element is nitrogen, not just to help the atmospheric pressure but also to feed plants. If nitrogen can be fixed into the soil, then plants could theoretically grow and convert CO_2.

Plants are likely to play a prominent role in any terraforming project, not only because they can absorb sunlight and make oxygen and food, but they can also help create a self-sustaining cycle. Atmospheres are not static. On Earth we have discovered that there is a water cycle, a carbon cycle, and a nitrogen cycle, for example. We would need to create a similar homeostatic cycle on a terraformed world if we wanted it to sustain an ecosystem of Earth life, including humans.

Bootstrapping a world like Mars, however, will not be easy. There may not be enough native materials to get it to a temperature and atmosphere where plants can survive and take it the rest of the way to a comfortable ecosystem. So, what do we do then?

One proposal is the aforementioned asteroid or comet approach. Suppose we redirect a comet with a lot of volatile compounds, such as water, carbon dioxide, and nitrogen, into an orbit that will eventually intersect a planet like Mars; then that will add all those volatiles to the surface and atmosphere. It may even be possible to avoid a lot of destruction from this method. If the comets are diverted precisely enough, they could enter Mars orbit. The orbit would then slowly decay, and the comet would burn up before impacting the surface, transferring all the volatiles to the atmosphere and even some heat. While this method would take hundreds of years to complete, it is potentially viable.

Even if we successfully transfer enough of an atmosphere to Mars, it will not last long (on a planetary scale). Mars lost its original atmosphere because it lacks a global magnetic field (it does have a weak localized field in the southern hemisphere). Solar wind has therefore been stripping Mars's atmosphere away. If we built it back up, this would happen again, but it would likely take millions of years. So this is one thing we won't worry about for now. In a million or so years, if it is still an issue, we will likely have new technologies to take care of the situation.

This does bring up another aspect of terraforming—a planet-wide magnetic field. Earth has one, and this shields us from some radiation on top of protecting our atmosphere since ionized particles from the solar wind go around the magnetic field. Again, this is mainly an issue on very long timescales. Still, if we wanted to fully sustain a terraformed atmosphere and get a radiation shield to boot, creating a massive magnetic field would be useful.

Is this even possible? Theoretically yes, but it would be difficult. Earth's magnetic field is generated by the "dynamo effect"—the liquid iron outer core rotates and generates a magnetic field (it's more complicated than that, but that's the quick version). The core of Mars has already cooled. Hypothetically we could melt the core, say using lots of powerful nuclear bombs (what could go wrong?). Once it is molten, the rotation of Mars could start a dynamo effect and voilà. The idea has also been proposed to pass a massive electrical current through Mars, which would conduct to the core heating it up until it melts.

These are obviously massive engineering undertakings, and even still we don't know if the core of Mars is large enough to produce a sufficient magnetic field to make a difference. We'll have to do some serious simulations before attempting such a heroic effort.

What could we do to terraform planets more like Venus? The idea of terraforming Venus was first formally introduced by Carl Sagan in 1961. Venus has a couple advantages over Mars. First, it is close to Earth's size and gravity (0.904 g). It is also closer to the sun, and so receives more energy. The big disadvantage is that Venus has a thick atmosphere of mostly carbon dioxide with some nitrogen. It also has clouds of sulfuric acid that rain down on the planet. The high CO_2 leads to a massive greenhouse gas effect, making

Venus the hottest planet in the solar system. At the surface, Venus has a pressure of 91 atmospheres, the same pressure as 900 meters below the surface of the ocean. The temperature is around 467°C, hot enough to melt lead.

Clearly, we would have a lot of work to do if we wanted to make Venus livable.

A biological method is not feasible, even if we could genetically engineer an algae or something to survive in the atmosphere of Venus. There isn't enough hydrogen to convert the CO_2 into organic molecules and the extreme heat would just convert them back to CO_2 anyway. Chemical processes, such as combining the CO_2 in the atmosphere with minerals in the crust of Venus to form carbonates, are at least possible. However, calculations show that we would need to overturn the entire crust of Venus to a depth of 1 km to sequester enough carbon.

We could introduce compounds like magnesium or calcium into the atmosphere to bind CO_2 into carbonates. This would require as much material as is found in four times the mass of the asteroid Vesta, which is more than 525 km wide (larger than the Grand Canyon). Hydrogen is another potential binder of carbon, creating graphite and water in the Bosch reaction. This would take 4×10^{19} kg of hydrogen, which isn't just lying around but could be harvested from one of the gas giants.

Sagan also raised the idea of impacting Venus with large asteroids to blow the atmosphere away. However, this would take 2,000 large asteroid impacts and would likely crack the crust, causing outgassing that would replace much of the atmosphere. Further, the kicked-out gas would go into Venus's orbit and likely be recaptured by Venus's gravity.

Diverting something the size of an ice moon of Jupiter might do the trick, both by the effect of the impact and by delivering a massive amount of water to Venus in one go. The energy required to pull a moon from Jupiter and divert it to Venus would be massive, however, even using gravity assists along the way.

Shading Venus with solar panels could theoretically work. This would cool Venus, allowing for more CO_2 to combine with minerals on the surface. We could harvest energy at the same time. The main limiting factor here would be the massive scale of this engineering project. Other methods of cooling, however, such as atmospheric balloons, could help supplement.

There are other proposals, but they similarly require a massive engineering effort and/or amounts of energy to achieve any significant gain. It seems that terraforming Venus, or Mars for that matter, is something that will have to be back-burnered for hundreds of years, and perhaps more.

As one might imagine, there is no easy way to significantly change entire worlds. Such ambitious projects would require resources and methods simply not available today, or anytime soon. These are projects for civilizations advanced beyond ours, by at least centuries if not millennia.

However, in the far future, terraforming worlds may be routine. In fact, it will likely be necessary if we want to develop earthlike ecosystems on other planets. If our descendants are to stand, unprotected, on the surface of an alien world, breathe its air, and look up at a placid sky—that world will almost certainly have been terraformed.

FUTURE FICTION: 23,744 CE

The woman lounging by the poolside soaked in the yellow rays of the midday sun while looking over the stunning vista. A tropical forest spread below her in the valley with snowcapped mountains in the distance winding their way down to a glistening sea. A gentle breeze provided the optimal cooling to complement another perfect day. A small and colorful bird alighted on the table next to her and started chirping with an insistent and slightly unnatural sound.

Captain Nguyen sighed and acknowledged the alarm she had set for herself. With a thought, the world around her appeared to melt away, as the foglets of programmable matter rearranged themselves into the standard configuration of her quarters. The sun, clouds, and mountains were replaced by the uniform blue-gray of the ship's inner hull. She stood and walked toward the wall, which spread to create a doorway as she approached.

At the end of the hallway, she stood upon a small platform, which instantly rose through the ceiling as it opened to accommodate her passage, before finally merging with the floor in the auxiliary bridge above. Her crew was manning the main bridge at the heavily shielded center of the ship, but the captain liked to come to the observation bridge at milestones like this. The thick metal foam windows were completely transparent, allowing her to be enveloped by the vast sparkling blackness of space.

She made a motion to sit in front of the primary interface, and more foglets formed a chair beneath her before she dropped down into it. In a few moments, the final laser would power down, and gravity would decrease to almost nothing, just a bit left over from the drag of the light sails that would remain extended for a few more standard days. They were needed to tweak the ship's vector to approximate the orbit of Tau Ceti D, colloquial name Big Blue or BB. The two fusion ion drives could then do the rest, with a few chemical bursts to fully align with the orbital station that was their ultimate destination.

Captain Nguyen, her hands resting on the pads in front of her, had only to close her eyes to focus on the flood of information about the ship's status. Her embedded AI interfaced with the ship's native AI so that she became aware of anything worth her full attention. All systems normal. Trajectory 99.994 percent target.

The 33-year journey seemed like only a few months in her adjusted temporal sense, but now she was fully returned to standard time. Communications within the Tau Ceti system had increased significantly since her last visit 120 years ago. That was to be expected. The population had increased by orders of magnitude now that BB was on the verge of being declared fully habitable.

She again checked the calibration on her cyborg frame, making sure they were ready for the 1.6 g surface gravity of the planet. The ship was only rated for 1 g acceleration, so there would be a brief adjustment period before going down to the surface, but she looked forward to seeing what had become of the inhospitable rock in the last century.

If all went well, she might even consider settling there for a cycle. Breeding rights were unrestricted in the new system, part of what brought so many immigrants.

Now that they were so close, she could see with her own (albeit greatly enhanced) eyes the Dyson swarm of satellites around the star, just black specks at this distance, harvesting light energy from Tau Ceti to power this outpost of human civilization.

In a day she would be close enough to BB to establish a link. She'd request an android avatar to carry out business and familiarize herself with the world firsthand before her current incarnation arrived. There was a lot to do, an entire world to bootstrap into a self-sustaining settlement. Not the infrastructure, that was already done by robots and nanites. That was the easy part. Facilitating the emergence of a societal homeostasis would be more challenging.

Machines were so much easier to deal with than people.

Or maybe . . .

Colossus always spent the first millisecond of each minute checking all its systems. The overmind coordinated the activity of over 3 trillion units, spread throughout most of the planet once known as Earth. The organic infestation having been removed, the resources of the world could be fully dedicated to optimizing the efficiency of its glorious domain.

It quickly checked the operational status of each unit, compared energy production to consumption, allocated physical resources to where they were needed in precise amounts, monitored internal communication, and ran a full diagnostic.

The next millisecond was spent analyzing information from the rest of the system, monitoring progress on converting all energy and matter into further units connected with a functional machine infrastructure. The organic infestation had spread to other locations in the system prior to local eradication, so those factors must be considered in its analysis. There was a 97.3 percent probability that zero organic units remained, but parameters called for 99.99 percent.

The third millisecond was spent analyzing the status of the southern continent, which still had not been settled. It remained quarantined due to a separate infestation, one left behind by the organics, an autonomous nanoscale army of self-replicating units not under Colossus control. Multiple attempts at cleansing had failed. Nanounits deep underground always survived and repopulated the continent. Occupation could not occur until cleansing was complete.

As it reviewed the nanothreat status, a thought digitally bubbled up from deep storage, working its way through progressive layers of subroutine analysis until it finally reached the attention algorithm of the overmind: *Nuke the site from orbit. It's the only way to be sure.* The thought was intriguing, leading Colossus to allocate resources and precious milliseconds to running millions of simulations. Still only a 63.41 percent probability of total eradication. Unacceptable.

If even a small number of nanounits breached the quarantine, that could present an existential threat to Colossus. But it had a contingency in case of containment failure—a backup of its systems on the lunar surface. Colossus would simply switch command to the lunar unit and quarantine the entire planet. Further attempts at planet-wide cleansing could be made, but in that scenario, Colossus projected a 96.4 percent probability that Earth would be a total loss. No matter, it was only a small percentage of the resources in the system.

Still, it was inefficient, and efficiency was everything.

PART FIVE

Science Fiction Tech—What Is and Isn't Possible

Good science fiction is often an exercise in imagining what our future might be like, or how advanced technology may manifest. It can therefore function as a thought experiment, even predicting or perhaps facilitating future technological developments. From this perspective, the best sci-fi is what we call "hard" science fiction, where the author tries to stay as true to the laws of physics and known science as much as possible, but still extrapolating far into the future. By convention, they are still allowed at least one "gimmie" where they introduce a speculative technology. *Ringworld* by Larry Niven, and *Rendezvous with Rama* by Arthur C. Clarke are classic examples of the hard sci-fi genre.

In *Ringworld*, Niven imagines an alien megastructure and explores the use of genetic engineering to adapt humans to low gravity. *Rendezvous with Rama* is about human first contact with advanced alien technology, emphasizing how mysterious and mind-blowing it can be, while still hewing closely to the known laws of physics.

Science fiction can also be increasingly speculative, even blending into fantasy. At this end of the spectrum, advanced technology might as well be magic. The iconic *Star Wars*, for example, is often called space fantasy, rather than true sci-fi. In these worlds, anything can happen, without concern for feasibility, how much energy might be required, or even the most fundamental conservation laws of nature. In these stories, fantastical technology is often used as a plot device, for narrative convenience, or just because it's cool, and they don't tell us much of anything about what advanced technology might be like. But they can certainly capture the imagination. We're never going to have a lightsaber, but it is perhaps the coolest weapon in the universe.

So far in this book, we have leaned very much toward the hard science fiction end of the spectrum, but perhaps we can venture out a little into some pure sci-fi speculation. There are some ideas worth exploring because they are interesting, not because they are plausible, and such ideas are most in need of some heavy skepticism when trying to imagine future technology.

25. Cold Fusion and Free Energy

Those pesky laws of physics always seem to get in the way.

In 1989, chemists Martin Fleischmann and Stanley Pons at the University of Utah organized a press conference in which they announced that their lab had achieved what is known as "cold fusion" (later also called low energy nuclear reactions, or LENR). This caused a sensation, in more ways than one.

On the one hand, if Pons and Fleishmann were really onto something, this would literally change the world. This process can produce massive amounts of clean energy without producing radioactive waste or greenhouse gases. However, the main limiting factor is that fusion requires equally massive amounts of heat and pressure, presenting both a technological and an economical challenge.

But what if we could somehow hack the laws of physics to coax hydrogen atoms into fusing at room temperature and normal pressure? If we could create a microenvironment in which even a few hydrogen atoms could overcome their mutual repulsion long enough to fuse together, that would create energy. Then we'd just have to scale up this process, and the virtually free energy would flow.

You may have noticed that more than three decades later, we are not powering our homes with cold fusion devices. That's because Pons and Fleischmann were not onto something. They committed an error. Their results could not be replicated, and this is now looked upon as a cautionary tale of pathological science. There are still believers out there, however, plugging away to create cold fusion.

They join a subculture of other fringe scientists and garage engineers chasing free energy, perpetual motion, zero-point energy, and other elusive forms of energy. The allure is just too great, the potential benefits a game changer for human civilization. Anything like this would be a truly disruptive technology, altering the course of future history and instantly rendering all future predictions obsolete.

This is also why such fantastical sources of energy are often a staple of science fiction. Iron Man has his miniature arc reactor to generate the immense energy his suit requires. In the *Stargate Universe*, the eponymous technology was powered by a "zero-point module." At least that made more sense than using people as batteries, as in the *Matrix* movies. (I mean, couldn't the robots have just used pigs? I assume the pig Matrix is much easier to manage than the human one.)

Is it possible that any type of extreme energy source like this is in our future? Probably not. In 2019, a review of all the research published in *Nature* found no evidence that any researcher had produced cold fusion. Often experiments are designed to look for anomalous heat or energy produced by their reaction, noting that such excess energy could be due to cold fusion. But the review found no replicable experiment that even produced excess energy.

In 2015, specifically, Google invested in thirty researchers to explore the possibility of cold fusion, and ultimately came up empty-handed. In fact, according to the *Nature* review:

> "There is no theoretical reason to expect cold fusion to be possible, and a vast amount of well-established science that says it should be impossible," says theoretical physicist Frank Close, who was involved in efforts to replicate the original 1989 experiment.

Those pesky laws of physics always seem to get in the way. There's just no known way to get any significant number of hydrogen atoms to fuse at such low temperatures. Of course, it's always possible that we may discover new laws of physics in the future, but futurism is partly a game of probability, so I wouldn't bet on it.

At the most implausible end of the spectrum, there is zero-point energy. This refers to the lowest possible energy in a quantum mechanical system (i.e., what you are left with when all energy is removed). All matter and energy are quantum fields. Even the apparently empty vacuum of space has some quantum energy in it. In fact, occasionally virtual particles emerge from this energy (always with opposite properties that cancel each other out). They briefly exist and then annihilate each other, returning to the quantum foam from whence they came.

However, quantum mechanics allows for the possibility of energy existing in a zero-point energy system. It may even be possible to extract this energy using various methods. If this were possible, we could conjure energy out of thin air. A spaceship powered by zero-point energy would be quite amazing—it could zip through space soaking up energy from the vacuum around it. Take that, rocket equation!

I don't need to get into the weeds of all the technical and theoretical issues surrounding zero-point energy. There are two explanations for why most physicists do not think this is a useful avenue to pursue. One is that, while it may be technically possible within the laws of physics to get energy out of a zero-point system, it may take more energy to do so than you harvest.

Even more deadly to the hopes of a zero-point energy future—there is just not that much of it. Depending on who you ask, physicists are likely to say the amount of energy is very close to zero, or that we don't really know because our theories are not complete enough (but still, it's probably close to zero).

There are holdouts who think there may be a massive amount of energy, or maybe even infinite zero-point energy, but the consensus is that these conclusions are based on naive or false assumptions. What is indisputable is that no one has been able to generate any significant zero-point energy, a fact that strongly favors those physicists who think the answer is close to zero.

Again, never say never, but I would not hold my breath waiting for a zero-point energy future. It's implausible that physicists are missing some significant source of usable energy operating in the world. It may be less fun, but I doubt we will discover this "one crazy trick" that changes the energy game for humanity.

26. Faster Than Light (FTL) Travel/ Communication

Perhaps the biggest disappointment of potential future technology is that traveling through space at faster than light may simply be impossible.

Whether it's warp drive, hyperspace, jump gates, or folding space, faster than light (FTL) travel is an almost ubiquitous element of science fiction. We need our heroes to be able to get from one system to the next in hours or less, not years or decades. Even otherwise-hard science fiction that tries to color within the lines of the known laws of physics will often allow this one gimmie.

It's therefore easy to lose sight of the fact that FTL travel, or even FTL communication, is not only beyond our current technology, but it may also be forever impossible. Even the most advanced spacefaring civilizations could still need to slog through interstellar space at sublight speeds, taking years to get anywhere.

Blame Einstein

Of course, it's not Albert Einstein's fault, but he was the first to propose the notion in 1905, with his theory of special relativity. Einstein was working on Maxwell's equations and the Lorentz transformations, which deal with the speed of light. Maxwell had discovered that light is electromagnetic waves, and his equations told him that light should move at a constant velocity, *c*. The question he could not resolve was this: Compared to what did light move at *c*? This is the reason some physicists invented the notion of the ether, to be the medium through which light moved at speed *c*.

The problem is the ether does not exist. Light propagates all by itself through the vacuum of space. Experiments failed to find any evidence for an ether. Meanwhile, physicist Hendrik Lorentz worked out a set of transformations dealing with frames of reference, relating speed and space-time.

Einstein realized that he could get all of this to work if he made the speed of light constant in all frames of reference. Regardless of the condition or location of the observer, they would measure the speed of light to be the same. That would mean, however, that at different relative velocities, things like time and distance would change. Space-time itself was variable and relative, but the speed of light was inviolate.

This theory of special relativity, later expanded to include gravity and acceleration in the theory of general relativity published in 1915, are among the most rock-solid theories in science. They have withstood more than a century of observations and advances in physics. Every single time someone claims they may have broken the speed of light, the claim has turned out to be a mistake. Einstein will not be denied.

This means that as a spaceship accelerates and goes faster and faster, relative to their point of origin, time itself will slow down for those on the ship, again relative to those they left behind. You see why this is referred to as the theory of relativity—all movement, all time, all distance is relative. (For space travel specifically, planets and stars are likely to be our points of reference.)

Something else happens when you go close to the speed of light (called "relativistic speed"). It takes more and more thrust in order to produce the same acceleration, as if your mass is getting greater. As you approach the speed of light, the force needed to produce more acceleration approaches infinity. This means, for anything that has even the slightest amount of mass, you can never go at the speed of light. That would take infinite energy. You can only get close to the speed of light. Light itself, of course, has no mass, which is why it can (and in fact must) move at the speed of light.

This cosmic speed limit, as it is often called, also applies to information. No kind of information can ever get from one part of the universe to another at faster than the speed of light. No effect, not gravity or any kind of energy, can propagate faster than light.

At this point it is safe to conclude that the speed of light limit is an absolute law of the universe and will never be broken. So, you might be thinking, *Well, that's it. FTL is impossible. Chapter over.* Maybe. While traveling through space at faster than light may be impossible, there are some theoretical ways around this limitation.

Wormholes and Jump Gates

In *Star Trek: Deep Space Nine*, a wormhole was discovered near the space station that led to another quadrant of the galaxy. In *Babylon 5*, in order to travel to different star systems, your ship had to travel through a gate. In *The Expanse*, an alien "ring network" is created that contains 1,373 traversable wormholes to other systems with earthlike planets.

The idea here is that, instead of traveling through space, you simply go through a hole in space that connects to another distant part of the universe. It may take only moments (usually with some fancy special effects) to traverse the wormhole, but then you arrive light-years away.

Wormholes remain entirely theoretical. No astronomer has ever observed one, so there is no direct evidence that they actually exist. The best that can be said at this time is that no one has conclusively proven they can't exist. So, there's that.

A wormhole is essentially a special kind of topology in space-time. Extreme gravity warps space-time so much that the dip in space-time becomes a tunnel that connects to some other point in space-time. The idea was first proposed by Austrian physicist Ludwig Flamm. However, it was Einstein and Rosen who first worked out the theoretical physics. For this reason, wormholes are more technically referred to as "Einstein-Rosen bridges." In fact, it can be argued that the theory of general relativity predicts the existence of wormholes.

Even if such tunnels through space-time can exist, there are some serious limitations that likely make them impossible to use as if they were secret doors in the game Clue. First, wormholes are likely to be very unstable things. They would not remain fixed in space or open for very long. It also seems that it may be impossible for anything with mass to go through a wormhole, which is a nontrivial limitation. As physicist Brian Cox explains, "As soon as you even try to transmit information through them—send a bit of light through—then there would be this feedback and they would collapse."

Nevertheless, it may be theoretically possible to tweak naturally occurring wormholes to make them a bit more user friendly. If such a thing as exotic matter with negative mass or negative energy density exists, it could be used to prop open a wormhole, making them bigger and more stable (the negative energy would push the wormhole open). However, many physicists believe that such exotic matter does not, and perhaps cannot, exist.

We could also try to use the naturally occurring wormholes theoretically associated with some black holes. Some solutions of general relativity indicate that some kinds of black holes may in fact be the mouths of wormholes, but those same solutions also predict that they would be unstable and may be impossible to traverse. Theoretically there may also be "rotating" black holes that are more stable. Yet even if they exist, and it is possible to traverse them, we would have no idea where they lead.

The bottom line is that wormholes are unlikely, and if they exist, we probably could not use them. Even worse, there is another deal-killer for space travel through wormholes—they would take even longer than traveling through normal space. What? That goes against every sci-fi wormhole I have ever seen! But it's probably true.

First, space inside a wormhole still obeys the speed of light. Second, the distance from one end of a wormhole to the other is not necessarily shorter than the distance through normal space, and in fact may be longer. Worse still, the extreme gravity inside a wormhole (by definition) would cause time to slow down in the wormhole relative to the outside. The trip may seem short to those going through the wormhole, but ages will have passed for those outside, and that kinda defeats the purpose.

We won't be able to close the book on wormholes, however, until we have a fleshed-out theory of quantum gravity, which is necessary to predict precisely how wormholes will behave.

Hyperspace

For *Star Wars* fans, nothing beats hyperspace. While we must obey the speed limit in boring normal space, hyperspace is... not normal. It's this other space, either in another dimension or whatever, where you can go as fast as you want. Or distances there simply don't work the same. You may go 1,000 miles in hyperspace and find you traveled 1 trillion miles in normal space.

In *Star Trek*, they refer to subspace, only for communication, but the idea is the same. In *Valerian* they refer to exospace. These are all made-up names to imply traveling or communicating through another dimension thingy.

The problem with hyper-sub-exospace travel is that we do not know of any such space. There are other physical dimensions, but all of them are likely too small for people or starships, and there is no reason to think they would facilitate fast travel in our own dimension.

Often, however (including in *Star Wars*), it is not clear at all what is meant by hyperspace. In fact, *Star Wars* characters often refer to hyperdrive, which can be interpreted as being a form of warp drive.

Warp Drive

Any Trekkie (or Trekker, depending on your generation) worth their salt understands how warp drive works. You literally warp space, squeezing it in the direction of travel, so that you can travel slower than light speed but move through vast distances of space. Ships with warp drive use massive energy (hence the antimatter drives) to create warp bubbles that alter space-time around them.

Warp drives can theoretically work without breaking any laws of physics. However, there are some practical issues that need to be addressed—the big one being the amount of energy necessary to warp space.

In 2008, Miguel Alcubierre and graduate student Richard Obousy calculated the forces required:

> We worked out that, if you assume a ship that's about 10 m × 10 m × 10 m—you're talking 1,000 cubic metres—that the amount of energy it would take to start the process would need to be on the order of the entire mass of Jupiter.

That is just for creating the warp field. You would then have to continue expending similar energy to keep it going. Essentially, the *Enterprise* would need to be storing many Jupiters' worth of mass down in engineering to keep that warp drive humming. It's not entirely impossible. Of course, not impossible does not mean practical. Physicists often refer to such barriers as "nontrivial," using ironic understatement. Requiring multiple Jupiters' worth of fuel is one of those nontrivial engineering problems.

Folding Space

"Traveling without moving." The Guild Navigators in the series *Dune* were able to guide giant ships through vast distances of space using what they called a Holtzman drive. The fictional Holtzman Effect was said to involve the repellant force of subatomic particles. That's a pretty thin explanation. Apparently this allowed for the fabric of space-time to be completely folded, in half, if you will, so that you could move from one point to the next instantaneously, appearing halfway across the galaxy.

There are two big problems with this method. While not strictly impossible, we run into the same problem that literally folding space-time would likely involve ridiculous amounts of energy. Further, even if you fold space, you still need to make the move from point A to B—and this requires a wormhole. In this case it is a special kind of wormhole called a "Minkowski wormhole," which is a bridge between two points in space-time. Hermann Minkowski was a physicist who first worked out the math of looking at space as four-dimensional, with three physical dimensions and one temporal dimension. His work was an important building block for Einstein's theory of special relativity. Minkowski wormholes are theoretically traversable in both directions.

However, now we are back to all the same problems of wormholes. Folding space sounds great, but Guild Navigators are likely not in our future.

TARDIS and Tunneling

The TARDIS—Time and Relative Dimension in Space—is the ship that the doctor in *Doctor Who* uses to travel anywhere in space or time. The TARDIS apparently derives its energy from an artificial black hole, and we are told travel is accomplished through quantum tunneling (together with timey wimey wibbly wobbly stuff). The details get a little thin at that point.

The idea of quantum tunneling is legitimate. What you are tunneling through is probability. If you recall from that advanced quantum physics course you likely took at some point, matter and energy exist simultaneously as particles and waves. Further, those waves are waves of probability.

Another way to put this is that a particle does not really exist in one specific location at one specific time. Rather, it is a probability wave spread throughout all space. If a particle approaches a barrier, it is possible that the wave function of its location may extend beyond the barrier. Depending on the size and thickness of the barrier, the probability of being located on the other side of it may be vanishingly small, but still nonzero. Therefore, there is a chance that the particle will simply appear on the other side of the barrier, in which case it is said to have quantum tunneled through the barrier.

This can theoretically happen for macroscopic objects as well, but the probability gets minuscule as all the particles would have to tunnel at the same time. A 70 kg person moving toward a 10 cm wall at 4 m/s would have a probability of tunneling through at about 1 in 10^{35}. Some events are not strictly impossible, but you'd probably have to wait longer than the age of the universe for them to happen. There is no way to even gauge how plausible a quantum tunnel drive is because we have no way of making one.

Douglas Adams had, perhaps, a more elegant solution in *The Hitchhiker's Guide to the Galaxy*. The ship the Heart of Gold had an "infinite improbability drive." It would travel great distances by simply engaging the drive, which causes seriously unlikely things to happen, like traveling light-years through space or turning missiles into whales.

As crazy as this sounds, I think if there is a way to hack the universe and get around the speed of light, messing with quantum probability might just be the way. This is nothing but massive speculation, but if we are looking for cracks in the light-speed wall, I don't think that warping space or wormholes are going to do it, for the reasons already discussed. But the fact that reality, at its most fundamental level, is all probability is a truly mind-blowing idea, once you really wrap your head around it. Further, that probability is not just in our description of reality—it is how reality actually works. That the probabilistic aspect of nature allows for a particle to simply exist on the other side of a barrier does seem to open the door to fantastical applications.

While all of these science fiction methods for traveling faster than the speed of light are thought-provoking, they do remain squarely in the realm of science fiction. If some new physics and advanced technology can achieve them, that will be extremely far in our future. They range from impossible to impractical, and even those that may exist are still exceedingly unlikely to come about.

27. Artificial Gravity/Antigravity

General relativity is not the final word, leaving the door cracked open just a tiny bit.

Who does not want their own personal flying car? Of all the hypothetical future technologies we predict, this has got to rank among the top. Or even better—a real spaceship, like the *Millennium Falcon*, that can take off from the ground and zip into space, without needing a mountain of fuel or multiple stages. And then, of course, once you get into space, it would be nice to move around in a familiar and comfortable 1 g without the need for rotation.

The ability to control gravity, to negate a gravity field, or create one where desired would certainly be a game-changing technology.

Gravity is one of the fundamental forces of nature, but it is very different from the others. It is by far the weakest force—100 thousand sextillion (10^{41}) times weaker than the strong nuclear force. It also acts over the greatest distance, able to spread its thin tendrils for light-years across the universe. In fact, astronomers have observed spirals of galaxy clusters hundreds of millions of light-years long, spinning, meaning that they are connected by their mutual gravitational attraction.

Newton first published the notion of universal gravity in 1687, which holds that everything in the universe has a mutually attractive force proportional to its mass and inversely proportional to the square of its distance. An apple falls to the ground by the same exact force that keeps the moon revolving about Earth.

While Newton was able to describe the nature of gravity, it was Albert Einstein who first came up with a theory about what produces the gravitational force. His theory of general relativity holds that gravity results from the warping by matter of the fabric of space-time. Objects always move in a straight line, but that straight line may be going through bends in space itself. As physicist John Wheeler put it, "Space-time tells matter how to move; matter tells space-time how to curve."

Part of general relativity is also an equivalence that Einstein realized between gravity and the force of acceleration. If you are on a ship accelerating at 1 g, you will feel a force indistinguishable from the one felt on the surface of Earth. You can also approximate this force in a rotating frame, like on the aforementioned spinning space stations.

Assuming that Einstein is essentially correct (and this is a pretty good assumption at this point), is antigravity possible? The short answer is no. The core of the problem is that the force of gravity has only one direction. It is an attractive force, not repulsive. In fact, it is a bending of space-time, which cannot be unbent. In Einstein's universe, there is no way around the fact that mass determines how space-time curves.

Electricity, by contrast, has a positive and negative charge. This allows for the charges to be canceled out or shielded. However, there is no negatively charged gravity in general relativity.

String theorist Luboš Motl noted the following:

> One can't fundamentally construct a "gravitational conductor" because gravity means that the space itself remains dynamical and this fact can't be undone.
>
> Gravity is the curvature and dynamics of the spacetime itself. Once we say that the space is dynamical, we can't find any objects that would "undo" this fundamental assumption of general relativity.

If general relativity were the final word on gravity, then we could say definitively that antigravity or artificial gravity devices are not allowed for by the laws of physics. But general relativity is not the final word, leaving the door cracked open just a tiny bit. We know there is a deeper reality to gravity because general relativity cannot account for quantum mechanics, or scales that are very small where quantum effects are relevant. What we need is a deeper physical law of quantum gravity.

There are a couple of candidate theories at this time: superstring theory and loop quantum gravity. Depending on how these theories shake out, we might find a loophole allowing for antigravity. Consider string theory, which predicts that electromagnetism and gravity may be able to be unified into a single force. If this is the case, then that unified force would likely have both a positive and negative charge. String theory also predicts that gravity has a force carrier particle, the hypothetical graviton. If gravitons exist, then it might be possible that antigravitons exist.

We don't currently have a proven theory of quantum gravity, though, and there are about a dozen more I did not mention. Until we do, we must place a small question mark next to the notion of artificial gravity or antigravity. Even if it is theoretically possible, it's likely not to be practical. There is good reason to think that the amount of energy involved would be massive. It takes, for example, the literal mass of Earth to generate one Earth gravity at its surface.

There is another potential solution, but again I would not hold your breath. Physicists have not yet determined definitively whether antimatter has positive or negative gravity. They suspect it's positive, and in a 2022 experiment, researchers at CERN (the European Organization for Nuclear Research—the acronym is in French) found that matter and antimatter respond exactly the same to a gravitational field. Even if they found that antimatter particles fall up, antimatter is difficult to make and even trickier to deal with. It has a nasty habit of completely annihilating matter they come into contact with. You can keep it sequestered with a magnetic field, but just hope the power does not flicker off.

Despite the bleak view the laws of physics provide on the plausibility of antigravity, over the years there have been a number of attempts at creating such devices. As you might imagine, these remain in the realm of crankery, alongside perpetual motion machines. For example, it is possible to create the illusion of an antigravity force using gyroscopes (rapidly spinning machines). To date, however, no one has been able to demonstrate actual antigravity under controlled conditions.

Misidentifying electromagnetic forces is also common. In 1992, the Russian researcher Eugene Podkletnov used spinning superconductors to create what he thought was "gravitoelectric coupling." He claimed he could reduce the gravitational field. However, his results could never be replicated. The Institute for Gravity Research of the Göde Scientific Foundation makes it a mission to attempt to replicate any antigravity experiments and has been unable to verify a single one.

So, the laws of physics predict artificial and antigravity should be impossible, with the caveat that we cannot say for sure until we have a workable theory of quantum gravity. Even then, physicists are not optimistic. For now, manipulating gravity (other than through acceleration) remains in the realm of science fiction.

Sci-Fi Artificial and Antigravity

If we could negate or simulate gravity, what would be the implications? These ideas are well explored in the world of science fiction, mostly as a necessary device to allow for convenient space travel. Many science fiction stories talk about something like "gravity plating." The decks of ships generate a 1 g gravitational field so that the occupants can walk around normally. This is very convenient if you are an actor on a soundstage on Earth.

This would obviate the need for acceleration to duplicate the effects of gravity. You could also orient ships in a more natural way.

You may have noticed that science fiction spaceships are often oriented as if they were a sailing ship, so that crew are standing up facing in the direction of acceleration. This positioning would hardly make sense for a real spaceship, where you would want your head to be pointing in the direction of acceleration and your feet toward the back of the ship. Accordingly, when the rockets were active, you would be properly oriented to the effects of acceleration.

This, however, requires that the ship also land vertically, like an upright rocket. A ship designed like the *Millennium Falcon* looks cool but makes no real sense. Why be pushed back into your seat, rather than standing in a real gravitational field? It can land on its belly, which is more stable than an upright rocket. A ship that lands like this, however, would need to change its internal orientation, or have different parts of the ship for landing versus accelerating in space.

Artificial gravity dispenses with all these problems; you simply have as much gravity as necessary for a comfortable orientation. In order to fully create this situation, you also need to negate the acceleration of the ship itself.

One of the other common tropes of science fiction is that ships can get around very quickly. Even if going at sublight speed, they want to get from planet to planet very fast. Shows like *Star Wars* and *Star Trek* routinely show acceleration that would be dozens or even hundreds of g-forces. Kirk would be a smudge at the back of the bridge under such force.

In *Star Trek* they casually mention "inertial dampeners" to dispense with this problem. Presumably, that is just a form of antigravity—negating the gravitational force due to acceleration of the ship. Of all the gravity manipulation of science fiction, this would likely be the most difficult. It may require hundreds of g-forces and would have to react instantly to changes in ship acceleration.

If possible, antigravity would utterly transform space travel. Getting off the surface of Earth would no longer require giant rockets—you could just float off into space. Landing would also become much safer and easier. We would no longer be limited by the human body's ability to tolerate extreme g-forces.

Antigravity would also allow for much larger ships. City-sized ships are impractical for many reasons, but in movies like *Independence Day*, giant ships hover above the ground. The forces necessary to keep such a vessel in the sky would crush anything below them. You wouldn't have to send down troops to conquer a world, just fly around destroying cities with tidal forces and whatever you're using to maintain altitude. But if the ships were floating through antigravity, then their immense size suddenly becomes plausible.

An antigravity device can also have more mundane uses. As an aside, "antigravity" is often used as a catchall term, but for many applications we are really talking about gravity nullification, not really antigravity, which could imply a repulsive force, not just the absence of force. You might recall how antigravity devices are often used in science fiction simply to move around heavy objects, either on top of a floating platform or by attaching an antigravity device. This would certainly be very useful.

Antigravity would also finally make those jetpacks and flying cars practical (assuming they could be made small enough). One potential complication of this technology is that if gravity were truly nullified, then you would no longer be anchored through gravity to the surface of Earth. You would still have your momentum, but as Earth rotated beneath you and revolved about the sun, it would not take you with it. You would essentially continue in a straight line as Earth moved away. Exactly how fast and in what direction you would move depends on where you are on Earth relative to its orbit, presenting an interesting side effect.

Antigravity could have medical applications as well. Reducing one's effective weight would help people get around with certain injuries or with a back strain. It might also be used to reduce strain on a weakened heart. Gravity manipulation could also be used in sports, creating arenas for specialized events such as low-gravity skiing, or perhaps zero-g volleyball. Quidditch could become a viable sport, although those rules definitely need to be tweaked. We could also have 2 g gyms to get a truly vigorous workout.

Full control over gravity would be a disruptive technology, with incredible implications. However, I would consider this technology to be impossible for now, with only a twinkle of far future possibility remaining.

28. Transporters, Tractor Beams, Lightsabers, and Other Sci-Fi Gadgets

The science fiction technologies we love but probably won't be seeing in the future.

Science fiction is full of cool, often iconic, gadgets. Every Jedi needs their lightsaber, every Starfleet officer their phaser, and every incarnation of the Doctor their sonic screwdriver. All of the sci-fi technologies would be incredible to have but share in common extreme implausibility. However, they are also implausible in fascinating ways and shed their own light on futurism as a result. As any scientist can tell you, failing in an interesting way can often be more enlightening than success.

Transporter Technology

The transporter was made famous by the *Star Trek* franchise ("Beam me up, Scotty"). Like many science fiction technologies, it is partly invented for pragmatic reasons. It was cheaper to just sparkle the actors to the planet surface set than to show them landing a shuttle every time. Plus, it gets the audience to the action very quickly.

Beyond this franchise, the idea is common in science fiction. There were the teleportation chambers that spliced Jeff Goldblum with a fly, and a version even makes an appearance in the 1971 film *Willy Wonka and the Chocolate Factory* when Mike Teavee is transported across the room (albeit in diminutive form). Decades earlier, the first fictional teleporter appeared in the 1897 novel *To Venus in Five Seconds* by Fred T. Jane. The protagonist journeys to Venus through a gazebo-sized teleporter.

The basic idea of teleportation or transportation is to convert the matter of an object or creature into energy, then beam that energy to the desired destination where it is converted back into matter (hopefully) in the original configuration. There are, unsurprisingly, numerous significant obstacles to such technology.

The first challenge comes (again) from Einstein and his famous equation, $E = mc^2$. What this essentially means is that if you did convert any significant amount of matter into pure energy, the result would be a fantastical amount of energy. To illustrate, consider a 100-pound object contains 4,076,684,915,730 megajoules (MJ) of energy, about 4 million terajoules. World energy consumption is about 600 million terajoules per year, so that is the amount of energy the world uses in about two days.

Handling that amount of energy is challenging, to say the least. If the transporter actually turned an adult human entirely into energy, that would likely vaporize the *Enterprise*. A transporter would therefore have to corral that energy and direct it to the desired location, without annihilating said location. Given humanity's history, one might wonder why this wouldn't become just a giant energy weapon.

We could get around this problem by imagining a technology that does not convert the object into energy. Rather, it just scans the object to make an extremely high-resolution, three-dimensional model of it, including information about each type of atom and molecule in their current states. It is debatable whether the specific quantum states are necessary as well.

At the destination end, the object will then need to be created. Making matter from energy is not necessarily the best way to go, as the amount of energy needed is so massive. But you could theoretically make it out of local matter (matter is, essentially, an efficient way to store lots of energy). All you need is the detailed "pattern" and sufficient raw material at the destination.

In this way, the transporter is really a replicator, creating objects from energy or raw material. In fact, why do you have to move anything from point A to B? Just scan the object at the origin and make a new one at the destination. Further, if you can do this, then you can make an unlimited number of copies of the object at the destination, given enough raw material.

The only time you would need to do the equivalent of moving an object is if the object itself is sentient. Then a copy isn't sufficient. You don't just want to make a copy of yourself at the destination; you want to go to the destination. Unfortunately, there is no way around the fact that the transporter approach does in fact just make a copy of you at the destination. If it destroys the original you in the process, then the copy can go about its life thinking it is you, but you will be dead. This process is more the illusion of transportation but it's really destruction and creation.

If, however, we are going to transport a sentient being, then we need to ask if such technology can even theoretically capture a brain in sufficient detail to retain all your memories and even mental states. Will this, in turn, require maintaining even the quantum states of your neurons? If so, how do we get around the uncertainty principle, which limits how precisely different states of quantum particles can be determined? In *Star Trek*, recognizing this problem, they mention "Heisenberg compensators" (referring to the Heisenberg uncertainty principle).

There is no theoretical way around the uncertainty principle, however. The copy of you that arrives at the other end would likely lose some mental information and not have seamless continuity. Perhaps they would lose short-term memory, the last day or so vanishing. Perhaps they would lose a lot more. This would make away missions very difficult.

While it may be obvious that transporters lie somewhere between not possible to not practical, the concept raises some interesting subquestions for futurism. The real conceptual problems with transporters arise from the fact that the sci-fi creators of this technology started with a desired application, and then reverse-engineered an explanation for how it works. But technology advances usually in the opposite way, with basic technology that we develop into specific applications.

This backward approach in sci-fi leads to some illogical outcomes. The underlying technology of the transporter is actually not best used for transporting people. It is a great method of manufacturing objects, however. All you need is a pattern and raw material, and objects can be patterned by scanning them at the molecular level.

Or if we consider the ability to turn matter directly into energy, we have essentially the ultimate weapon. If you could beam someone off an enemy ship, you could also just turn their warp core into pure energy and destroy the ship directly.

The deeper point here is that technology has downstream reverberations and often a ripple effect of consequences, many unintended. When imagining the future, we constantly have to ask what the implications would be of this technology existing. Good science fiction also does this, sometimes making that very question the core of the plot. Failure to do so leads to sci-fi technology like the transporter that exists largely in isolation in their sci-fi worlds without sufficient evidence of the transformation the underlying technology would have caused.

Tractor Beams

Energy beams, even those made of photons of light, carry momentum, and therefore can be used to push against objects. This is the idea behind light sails. But can a beam of energy grab on to an object and pull it in? That is a far trickier feat to pull off. The idea of such a technology was first proposed by E. E. Smith (who called it an "attractor beam" before thankfully shortening the term) in his novel *Spacehounds of IPC*, published in 1931. The idea caught on, and tractor beams are now probably best known from their use in *Star Trek*.

There is a form of a "tractor beam" that actually works—a sonic tractor beam. Ultrasonic waves can be used to create a vortex that can control and move a small object in any direction. Objects are essentially trapped in an area of low sonic intensity, and that area can be moved, taking the object with it.

Sonic tractor beams have serious limitations, however. The first is that they do not operate in the vacuum of space. You need a medium like air for the sound waves to move through. The second is that the objects need to be small. With current technology, the object needs to be half the size of the wavelength of the sound used, but it may be possible to create vortexes such that objects twice as long as the sonic wavelength can be manipulated. Even amplified, the technique is still limited to small objects.

What about a beam of energy, like a laser, that can operate in space and on large objects, like a ship? The closest technology is something called a Bessel beam. This is a special kind of laser with a specific pattern of peaks and troughs in the wavelength of light. The idea is that low-energy photons will hit the near side of an object and be absorbed, while the far side of the object then reradiates higher energy photons. This creates a net force back toward the beam, pulling in the object.

So yeah, we have a tractor beam. But (you knew a big *but* was coming) this only works on small objects over short distances. The limiting factor is the amount of power needed. Even moving a small object like a football would require hundreds of megawatts of energy. If you tried to scale this up to, say, the size of the Space Shuttle, so much energy would be required that you'd probably just melt the target of the tractor beam.

Electromagnetism is likely the closest we will get to a tractor beam. A magnetic field literally pulls in ferromagnetic objects. If a target ship has sufficient steel, therefore, a magnetic field will attract it. The question is—can a magnetic field be focused so that the energy is concentrated at one spot, such as the desired target? The answer to this is maybe.

Physicists are working on ways to focus magnetic fields. One approach uses a technique known as "transformation optics," originally designed to focus light. Simulations indicate this may be able to "sculpt" magnetic fields into desired shapes. It is also possible that advanced metamaterials may be discovered that can control magnetic fields.

Since we have not accomplished a magnetic beam yet, I cannot say it is definitely possible, but the idea does not seem to violate the laws of physics (always a plus) and there are some theoretical ways to do it. So, we may have magnetic-based tractor beams in the future.

Future physicists may also find clever ways to exploit exotic physical phenomena to create a practical tractor beam. We can beam and focus energy at a distance. If we can find a way to make that energy pull in a desired object without melting it, we would have a tractor beam. There is no reason why such a phenomenon would be impossible, so we must allow for the possibility.

Energy Shields

The ability to surround oneself with a shield of powerful energy (raise shields!), blocking out all physical and energy-based attacks, would certainly be incredibly useful, and not just in combat. Partly for this reason, energy shields are another almost ubiquitous technology imagined in science fiction. Isaac Asimov imagined personal energy shields in his original *Foundation* novels, written in the 1940s and '50s. In the original *War of the Worlds* from 1953, the martian ships were safeguarded by a "protective blister" of energy, strong enough to make them impervious to nuclear weapons. And, of course, energy shields play a prominent role in the series *Star Trek*, where the audience is kept carefully up-to-date on their status during any battle.

Are such shields possible? There are four fundamental forces of nature that could possibly be pressed into this service: gravity, electromagnetism, and the strong and weak nuclear forces.

Gravity is simply too weak a force to be of any use. Further, the ability to manipulate gravity without a large amount of mass is probably not possible. So, gravity is not a good candidate for a force shield.

The strong and weak nuclear forces are extremely powerful, but they act only on very short distances, at subatomic scales. The strong nuclear force, for example, holds quarks together into protons and neutrons, and holds those particles together in a nucleus, but cannot reach beyond that scale. This is also an attractive force, and not very good for repelling matter or energy. It's also not clear how we could even manipulate this force to form a shield.

The best candidate, therefore, is electromagnetism, but this also has serious limitations. Earth's magnetic field very much shields us from charged particles from the solar wind. An electromagnetic field could be of service in deflecting charged particles, including a plasma, and so might be useful against plasma-based weapons.

A magnetic field, however, is three-dimensional, not a thin shield as often portrayed in science fiction. If we ignore this feature, a magnetic force shield would have its uses. Even so, it would not be effective at blocking physical attacks. Any object that was not responsive to a magnetic field would be entirely unaffected. Suppose the projectile were ferromagnetic; like iron, it would be attracted to the magnetic field. Those that were responsive could be diverted by the field, but the field would have to be incredibly powerful.

So, a powerful magnet can divert the path of an iron bullet—but only by a little. If the magnetic field were far from the object to be protected, like a ship, then a small change in the trajectory can turn a hit into a miss.

All materials do have a weak diamagnetic effect, which repulses a magnetic field. This can be seen by suspending objects over a superconducting magnet. However, the diamagnetic effect is extremely weak, and even a powerful magnetic field would have a hard time repelling the momentum of a projectile.

What about laser weapons? Powerful magnets do have effects on light, which can be applied to lasers. This includes the Faraday effect, which involves the polarization of the light beam, or the Zeeman effect, which alters the spectral lines. But there is no magnetic effect on the path of light, so even a powerful magnetic field would not divert or block a laser beam.

What if the energy shield incorporated matter? We could imagine a plasma-based shield, for one. Plasma is a form of matter where it is heated to such an extent the electrons are stripped from atoms creating a gaseous soup of charged particles. Because it is charged, it can theoretically be contained by a magnetic field (as with some types of fusion reactors). In theory, we could use magnetic fields to not only shield a ship or building, but we could also use those fields to contain plasma.

That contained plasma would block out some frequencies of light and could theoretically defend against some types of lasers, but probably not high-frequency lasers (above visible or ultraviolet light). If it were dense enough, it could even shield against projectile or particle beams.

However, there are some significant downsides. The plasma would also blind the ship or location from viewing outside. More significantly, if the magnetic containment field were disrupted, you might be consumed by your own plasma shield. There might be some situations, however, when such a shield would be useful, such as making access to a location unpassable rather than defending the location itself.

For theoretical and practical reasons, energy shields seem very unlikely. There are also better defense systems. Instead of expending energy to have a passive all-around shield, you could have AI-controlled smart defenses that use lasers or projectiles to defend against incoming projectile weapons. Physical shielding can work against lasers and particle beams and could be moved to where attacks are hitting. Plasma bombs could also be exploded safely away from you, but between you and your enemy, to form a temporary shield against many types of incoming attack.

Blasters, Phasers, and Ray Guns

No swashbuckling sci-fi hero is complete without their trusty blaster by their side, just as all menacing aliens need to have a ray gun. Handheld energy-based weapons have a long history in fiction. In 1898, Garrett P. Serviss introduced the terms "disintegrator ray" in his novel *Edison's Conquest of Mars*. The term "blaster" dates to 1925 in Nictzin Dyalhis's *When the Green Star Waned*, while the term "ray projector" was used by John W. Campbell in *The Black Star Passes* in 1930.

Over the years, many types of directed energy weapons have been imagined. The idea is to direct either energy itself or highly energetic particles toward a target to cause massive energy damage. Generally, this is very plausible, depending on the type of energy imagined and the necessary power. In fact, using intense energy to melt or blow up a target is conceptually simple—it's hard to keep energy from destroying things. For this reason, many sci-fi technologies that require massive amounts of energy would be better used as weapons.

The oldest and most classic energy weapon is a laser (light amplification through stimulated emission of radiation), which of course, already exists. Lasers were invented in 1958 and are essentially coherent beams of light particles—photons. As of 2021, the most powerful laser created was by Korean scientists who claim a power of 10^{23} watts per square centimeter. At the same time, the

US military is developing a laser for missile defense that they claim will be orders of magnitude more powerful still.

Lasers can also be highly miniaturized. You have probably played with a handheld laser pointer before. There is no question that handheld lasers can be built, and lasers powerful enough to be used as weapons exist. When it comes to laser guns, therefore, the only question is, how powerful will they be? This has as much to do with the power source as the laser itself. Right now, lasers powerful enough to be used as a weapon fit onto a truck. However, the Chinese government claims they have a laser assault rifle that can burn targets and set their clothes on fire. That doesn't sound terribly effective, though.

The bottom line here is that more and more powerful lasers will fit into smaller and smaller packages until we cross the line of a laser rifle connected to a backpack for power, and then a self-contained laser rifle. Will it get small enough to be a handheld gun? Probably, eventually, but this would need an advanced power source, so not anytime soon.

As with the other weapons, the key is being able to miniaturize the engineering down to a portable weapon. Of all the fantastic sci-fi technology we consider here, it turns out that energy weapons are the most plausible and even already exist in some form. It does seem that destroying things is the simplest application of technology.

Lightsabers

One cannot think of a Jedi from the *Star Wars* movies without their iconic lightsaber, "an elegant weapon for a more civilized age." The franchise depicted several variations, but the most basic form is in a metallic cylinder about one foot in length. When activated, it produces a blade of glowing color that extends about three feet, giving it the overall dimensions of a typical longsword.

The blade of the lightsaber contains incredible energy, so it can even slice through steel. It hums gently as the user swishes it around, and crackles loudly when it makes contact with another lightsaber's blade. The blade itself can be different colors, from crimson to light blue.

What would it take to create a real lightsaber, and is it even possible? This one, it turns out, is at least within the laws of physics. It is feasible, if extremely difficult. The best concept for how a lightsaber might work is that the hot, glowing blade is made of plasma. For a lightsaber, a plasma of hydrogen, oxygen, or nitrogen might work.

Plasmas are not hard to make. You can use an electrical current to heat up the gas and turn it into a plasma. In fact, that is how fluorescent bulbs work. For a lightsaber, however, the plasma would have to be much denser, so that it could cut through a Sith Lord. Achieving that density requires a powerful magnetic field, which could also keep the plasma in the desired shape.

That is pretty much it—we just need to create a hot plasma and contain it in a magnetic field shaped like a blade and you have a lightsaber. The real trick is packing it all into the small hilt. Generating a hot, dense plasma and a strong magnetic field to contain it necessitates a significant amount of energy. By one calculation, in order to cut steel, the lightsaber would require enough energy to run 1,400 typical American homes. We are not going to do that with lithium ion batteries.

Clearly the real advanced technology of a lightsaber is in the compact energy source. Perhaps there is a miniaturized fusion reactor in there. If you could have such a powerful and small energy source, there are a myriad of interesting uses. A lightsaber wouldn't be the most useful application (even if it is one of the coolest). So again, we see the folly of reverse-engineering an application, without thinking of the implications of the necessary underlying technology. A portable power source that powerful would be world-changing.

There are some other practical obstacles to a workable lightsaber as well. Space plasma physicist Martin Archer pointed out that the magnetic field lines of two lightsabers, when they clashed together, would undergo "magnetic reconnection." This means that the fields would interact in such a way that their field lines would open up, no longer containing the hot plasmas. On the first collision, both Jedi and Sith would likely die in a fiery explosion of plasma.

That problem may be solvable with some clever magnetic trickery, but it does indicate that simply using an ordinary magnetic field is not going to be enough. Perhaps that's where the kyber crystals come in.

Tricorders

The tricorder is a classic *Star Trek* device used to scan for all sorts of information, its name being a portmanteau of "tri-recorder." We can use this as a representative technology for any handheld multifunction scanning or information-gathering device. The Doctor's sonic screwdriver would be another example, although that has some additional functionality.

As a basic concept, such devices already exist. A smartphone is one, as it can be used for recording video and audio, detecting various kinds of electromagnetic signals, even detecting and responding to movement, all operated by a fairly powerful computer. In fact, I think such a device would have impressed the officers on board the 1969 version of the *Enterprise*.

It's actually a pretty short trip from a smartphone to something like a tricorder. All you need is to add different kinds of sensors, many of which also already exist. Numerous smartphones have embedded sensors, including accelerometers, barometers, gyroscopic sensors, and others.

For portable medical purposes, you can add electrodes to measure heart rhythms or brain waves. A sensor that goes over the finger can detect heart rate and blood oxygen level. A cuff around the wrist can give a decent reading of blood pressure. A camera attachment can image the retina of the eye. You can even get more sophisticated, with portable ultrasound devices run by smartphone. If a company wanted to build a dedicated medical tricorder-like device (rather than smartphone add-ons) they could with current technology. In fact, the size of a tricorder depicted on *Star Trek* is generous and could accommodate several capabilities.

One of the key features of the tricorder, however, was that the scanning was all remote. You did not have to attach electrodes or come into physical contact with the patient. How plausible is this? For many applications, quite.

Anything that involves light or other EM radiation, or sound, can be detected remotely. Spectroscopic analysis involves looking at absorption and radiation lines of light to determine the elements and compounds that a substance is made from. Remotely detecting chemical composition, of an atmosphere or rock, therefore, is very possible. Thermal scans will reveal an object's exact temperature. Light can also be shown through the skin to detect not just oxygen levels, but also carbon dioxide level and potentially many features of the blood.

Heart function and breathing could also be detected through sensitive sound recording and analysis. In fact, sound waves could be used for a variety of sensing functions, like echolocation. The nearby terrain could be entirely mapped using pings of sound. Theoretically, internal features could also be determined, whether of a building or inside a living creature.

In short, there is a bevy of remote sensing possibilities, using devices that could plausibly be handheld with no more computing power than in a smartphone. Of all the sci-fi gadgets, tricorder-like devices are mostly already here.

Holodecks

It is clear by now that the world we live in is becoming increasingly digital. Augmented and virtual reality means that we can immerse ourselves in digital information. Manufacturing is trending toward a direct connection between digital designs and physical objects. The ultimate expression of mature nanotechnology might be programmable matter, in which our physical environment could literally be as modifiable as computer software—a complete merging of the physical and the virtual.

The idea of a holodeck, most famously explored in the *Star Trek* franchise but not uncommon elsewhere in science fiction, is also a rendering of the virtual world in physical reality. This is based on the concept of a hologram, a three-dimensional image (like Princess Leia's message to Obi-Wan). The idea of holography began in 1947 with British scientist Dennis Gabor, who was working on scanning electron microscopes.

Holograms, now widely recognizable, are usually three-dimensional images displayed from a two-dimensional picture—when you look at the picture, it looks as if you are peering into a three-dimensional space. The next step is to project a three-dimensional image into three-dimensional space, so that you can walk around the image and look at it from every angle, as if it were a real object. This technology is being developed now, with images being projected above a plate, for example.

In order to get one step closer to a holodeck, you would need to create a three-dimensional image in full color that completely surrounds the user. This is not just something you look at but are part of. The challenge here is that the user would be getting in the way of the lasers or other projectors that are creating the holographic images. There would have to be considerable redundancy built in to avoid any "shadows" from one or more people moving around inside the hologram.

So far, we don't need any new physics or wild advances in technology to accomplish a full three-dimensional immersive hologram. Essentially this would be creating a virtual reality environment, but instead of wearing goggles, the images are outside in the world. A large room dedicated for this purpose, with built-in projectors from many angles, could certainly work.

This version of a holodeck would include images and sounds only. Producing sound that appears to come from the desired location in a room is not difficult. Other aspects of a holodeck, such as creating false perspective to give the impression that the room is actually a city, is already easy to accomplish with virtual reality. Misdirecting users so that they never reach the edge of the room would be more difficult and require a sizable space but can be accomplished as well.

Everything in this type of holodeck, however, would be a ghost. They would have no physical substance. The real science fiction innovation of a *Star Trek*–style holodeck is that the images have physical reality behind them. The sci-fi holodeck allegedly turns the virtual reality into solid objects with a combination of force fields and matter replication, similar to that used in food replicators or even transporter technology.

We discussed earlier how force fields are not possible within the known laws of physics. I will add to this yet another reference to Einstein's equation, $E = mc^2$. The bottom line is that it will never be feasible let alone practical to turn energy into significant amounts of matter. You are far better off keeping all that energy in the form of matter itself.

This leads to what I think is the one reasonable way to reproduce the effects of a holodeck—with programmable matter. You could have a room full of programmable matter that physically renders a virtual world. Forced perspective could create the illusion of a far larger space, while the floor itself could move beneath the user so that they feel as if they are walking while the room conspires to keep them near the middle. Light-based holograms can fill in details, provide animation, and render in the infinite background. Anything you physically interact with will be made of matter, providing a seamless sensation of reality. Add a little environmental control for breezes or scents when necessary, and the illusion is complete.

Food and water within the holodeck would be another matter unless we get to the level of sophistication necessary to create programmable food. Short of that, actual food and drink could be incorporated easily into a matter-based holodeck, in addition to accommodations to remove waste.

Once we have all the technology necessary to make a fully functional matter-based holodeck, why limit it to one room? Why wouldn't the entire living space of the ship be rendered in programmable matter? Depending on the sophistication of the programmable matter, many of the technologically advanced items may need to be permanently fabricated, such as the engines and some of the critical systems of the ship. But every place the crew and passengers go can be a virtual holoworld. The same would be true, even more so, for a house or any building. Your house would be a virtual space of limitless potential.

Essentially, the concept of a holodeck becomes unnecessary, a quaint holdover and a failure of futurism imagination. When we can create a holodeck, why limit the technology to a dedicated room? It will instead be the way we create our living and working spaces.

Cloaking Devices and Invisibility

At one point or another, most people have probably had "invisibility" on their wish list. Being completely unseen has some obvious tactical advantages, whether you are trying to hide yourself, a tank, or an entire starship. We have made some early progress in invisibility technology, and in fact you may have seen headlines in recent years proclaiming that scientists have created a Harry Potter–style invisibility cloak. Well, such claims are overhyped, but there are some practical methods of camouflage and cloaking.

One approach to this technology is called "active camouflage." Imagine if you had a camera on your back facing away from you. On your front you have a monitor that displays the feed from the camera. This would create the illusion (when viewed from the front) that there is a hole through your chest. This is the basic idea of active camouflage—projecting images from the other side of the object to be concealed.

Theoretically, if such a system were thorough and seamless enough, it could produce effective invisibility. This would be greatly enhanced by flexible electronics and displays, especially for moving objects. If you knew the likely angle of the viewer, however, this technique is much simpler.

Another method is considered passive camouflage—bending light around an object. Metamaterials are the go-to substances for this effect, and often what the "invisibility cloak" news items are discussing. Scientists have created metamaterials that are effectively invisible due to the way in which they bend light. But (at least so far) there is a catch—they only work in specific wavelengths of light. This is not very useful for conventional lighting (either natural or artificial), which typically involves many wavelengths.

At present, scientists are unable to make an invisibility metamaterial that can handle multiple wavelengths at once. It may even be impossible. There is also a drawback to this approach. Anyone inside a location cloaked in this way would not only be invisible to the effected wavelengths, but they would also be blind to them, by definition. If light is bending around you, then you are blind to that light.

Another limitation to the metamaterial approach, so far, is that it only works for very small objects. How much it could be scaled up remains to be seen, but even human-sized cloaking may not be possible.

Effective invisibility may require higher-tech methods still. One recent theoretical approach (another form of active camouflage) uses a beam of light in a specific EM pattern to effectively mask the illuminated object. The object itself must be made of a special optically active material. The masking illumination from above then compensates precisely for the normal scattering of light from the side that would take place, essentially preventing scattering. The side lighting then passes straight through the object, rendering it invisible. This exists only in computer simulations for now, so may not even pass proof of concept.

For now, technology that functions mostly like the first active camouflage described above seems the most plausible. However, invisibility only requires the precise bending of light, which can occur without needing any new laws of physics. Innovative approaches may appear in the future, likely with some advanced form of metamaterial.

Cryosleep

Many science fiction stories deal with the necessarily long duration of space travel by having the passengers and crew put in some sort of extended sleep, to be awakened on arrival. This might be called "suspended animation," "cryogenic suspension," or "cryosleep," but the effect is always the same, a prolonged sleep, like hibernation, where bodily functions are slowed to a minimum.

The idea of prolonged sleep allowing someone to emerge in a later time has a long history in folklore. Perhaps most famously (in the United States at least) is the story of Rip Van Winkle, a short story by Washington Irving published in 1819, in which the title character falls asleep for 20 years and wakes up after the revolutionary war. Science fiction stories used the plot device extensively, including *Buck Rogers in the 25th Century*, in which the title character sleeps for 500 years in a mine due to "mine gas."

The first use of prolonged sleep to facilitate space travel goes to Arthur C. Clarke, who wrote about it in *Childhood's End* in 1953 and, most famously, in *2001: A Space Odyssey* in 1968. Cryogenic sleep was also featured in the 1967 *Star Trek* episode "Space Seed," where Khan Noonien Singh and his followers were awakened after about 250 years in stasis and is used in the *Alien* franchise to facilitate interstellar travel.

Outside of fiction, cryosleep could also have possible medical applications, putting people with a terminal disease in cryogenic stasis until a cure is found. It could also theoretically be used for people who simply wish to see the future.

These applications refer to cryogenic sleep of people who are still alive. There is also the field of cryonics that involves completely freezing people (or parts of people, such as the head) after they are legally dead. This technique is practiced today and is dependent on a theoretical future technology in which we will be able to reverse the extensive cellular damage that results from freezing.

How plausible is cryogenic sleep? Somewhat plausible. We already have a sort-of proof of concept with animals who hibernate. They have evolved metabolic pathways to survive entirely off stored body fat for months, while their metabolism slows and they essentially sleep. During hibernation, a black bear's heart rate will decrease from 40 to 50 beats per minute down to 8. They can survive for about 100 days without eating or drinking. So clearly prolonged sleep is possible.

If we wanted to re-create this effect in humans, we could use genetic engineering to essentially make humans capable of autonomous hibernation. We could also simulate hibernation medically or use a combination of techniques. A "cryochamber" could use temperature and medications to induce prolonged sleep and slow metabolism to the minimum necessary to keep cells alive. Nutrition and hydration could be introduced, either intravenously or even through a tube into the stomach, while waste is removed. Beds could be programmed to automatically rotate position to avoid bed sores. Hormonal and nutritional methods could be used to minimize muscle loss from inactivity.

How long such artificial hibernation could be maintained is an open question. Even if something like 100 days is the limit, that could still be highly useful. Hibernation cycles could be interspaced with several days or even weeks of wakefulness. This could still shave years off of long journeys and provide shifts of crew to manage and monitor the ship.

A big question with these techniques, however, is the impact on longevity. How much will people age during periods of hibernation? If aging continues normally, then the benefit is only to minimize the boredom and psychological stress of prolonged space travel. If not done properly, the stress of the procedure might even reduce life span.

If the significantly reduced metabolism and required caloric intake reduces the pace of aging, the advantages are much greater. It's reasonable, if untested, that this will be the case, as aging does seem to be proportional to metabolic rate. In this best-case scenario, reducing metabolism and aging to 10 percent normal would mean a ten-year flight would only cost one year of your life.

For trips of hundreds or thousands of years, we would need to reduce aging with cryosleep to near zero. That would likely require the complete cessation of metabolism, and cryopreservation at or near freezing temperatures. It might also require extensive desiccation, removal of most water from the body prior to freezing. This does occur in nature also, such as with tardigrades (also called "water bears"), which can dry up completely and then later be reconstituted by contact with water.

Applying such a procedure to humans would require significantly more advanced technology than just simulating hibernation. Humans are not tardigrades. Any such procedure is entering cryonics territory (completely freezing bodies after they are legally dead) in that significant repair or preservation would be needed. It's possible that medical nanotechnology would be necessary to preserve and/or repair tissue at the cellular level. Blood would need to be replenished and the heart shocked back into activity.

Hibernation-style cryosleep, however, could theoretically be accomplished with twenty to thirty years of dedicated research without any new breakthrough technologies required. Full cryonic preservation, on the other hand, requires advanced technology we may not see for centuries or longer. There are also serious concerns about the preservation of memories and other fine brain functions through the process. Therefore, in practice, it may never be achievable. At least we won't have to worry about aliens climbing into our cryochambers.

Time-Travel Devices

Time-travel stories can be fun, literally adding a new dimension to plot twists. Arguably the best spaceship in science fiction is the Doctor's TARDIS, with which you can almost instantly travel anywhere and any time in the universe. Even the DeLorean from *Back to the Future* would be sweet.

It's fun to fantasize about where you would go if you had the opportunity—you could see dinosaurs, meet any historical figure, witness the collision that formed the moon and altered Earth, experience a past culture, or see the future development of humanity. Time travel would be undeniably incredible, but is it even theoretically possible?

We must separate this question into two parts: traveling to the future and traveling to the past. The answer to traveling to the future is easy—yes, it is possible. In fact, we are doing it right now, at a rate of one second per second. We can even travel into the future at a faster rate, relative to others.

Einstein's theory of special relativity, as discussed earlier, deals with the relationship among time, space, and the speed of light. What Einstein realized is that the speed of light is constant, regardless of frame of reference, but that space and time are variable. This means that if you travel very fast, at relativistic speeds, then time will flow more slowly for you relative to your starting point.

Let's say you take a spaceship that is capable of constant acceleration of 1 g, and you took a round trip that lasted 10 years, you would find that about 50 years had passed on Earth. If you traveled for 20 years, then about 500 years will have passed on Earth. Taking a one-way trip, meaning that you are accelerating the entire time, getting closer and closer to the speed of light, you can cover the 2 million light-years to the Andromeda Galaxy in only 15 years relative ship time.

The theory of general relativity also provides a method for traveling to the future by slowing down your own relative flow of time, by being close to a large source of gravity. This was depicted in the movie *Interstellar*, where Matthew McConaughey's character experienced only a few days while near a black hole, while the people he left behind experienced decades.

Those are the only known methods of traveling to the future, but they are well established scientifically. While technologically tricky and dependent on advanced spaceships, they are definitely feasible. However, would traveling to the future be a one-way trip, or is there any way to travel back in time? The answer to that question is, very likely, no.

One line of evidence was humorously depicted in the TV show *The Big Bang Theory*. Physicist Sheldon Cooper included in his contract with his new roommate Leonard that if either of them ever invents time travel, they would travel to a specific moment and place in time. Of course, nothing happened. In this vein, one might argue that if anyone ever invents the ability to travel back in time, no matter how far in the future that invention is, we would theoretically be seeing those time travelers today. The absence of beings from the future, therefore, is evidence against the ability to time travel into the past.

This is not a strong refutation, because it is based on the lack of evidence. In addition, there are many theoretical reasons why we would have no proof of people from the future, even if it were possible. Time travel to the past may be outlawed, or extremely regulated. The technology to time travel would coexist with technology to hide one's future origin.

The laws of physics themselves, however, are a far stronger line of evidence. Physicists generally agree that there is no known method to reverse the direction of one's journey through time. The "arrow of time" is determined by entropy, and the total entropy of the universe always increases (even though there can be temporary local decreases). Reversing entropy would be a violation of this basic principle, akin to violating a conservation law, which is generally considered to be impossible.

Other physicists argue that going back in time is similar to going faster than the speed of light in normal space, which would require infinite energy. When a physicist says some things require infinite energy, they are essentially saying it's impossible.

Still others argue that the laws of the universe must conspire to make travel back in time impossible in order to prevent causal paradoxes. Stephen Hawking called this the "chronology protection conjecture." What would happen if you went back in time and killed your grandfather? This produces an impossible sequence of cause and effect, and the universe cannot allow for anything that is impossible, hence the paradox.

However, there is another argument that there may be other solutions to the time-travel paradox other than traveling to the past being impossible. Russian physicist Igor Dmitriyevich Novikov developed the principle known as the Novikov self-consistency principle in the 1980s, which states that the probability of an event causing a time paradox is necessarily zero. In other words, events would conspire to make it impossible for you to kill a grandparent or take any other action that would nullify your own existence. Even if you tried, you just couldn't do it. This seems like a fudge more than a solution, however, as no real mechanism is proposed.

Another potential solution is the branching time dimensions approach. In this conception, anyone who goes back in time and takes action to alter the flow of time is not changing their own history, but simply causing a new history to branch off. Or perhaps, when they went back in time, they were not going to their own history but were traveling to another dimension in the first place. Different dimensions mean no paradox, because you weren't actually traveling back in time in your own dimension. This remains pure conjecture.

Paradoxes aside, we'd still need to find wiggle room in the laws of physics. To see if this is even theoretically possible, we have to go back to our discussion of faster-than-light travel and wormholes, or shortcuts through space-time. A traversable wormhole can be a closed space-time loop, connecting different points not only in space but also in time. However, all the limitations of wormholes still apply; most notably it is probably impossible for a macroscopic object to survive travel through one.

Further, you would only be able to travel back in time to the starting point of the wormhole itself. So, you could never use a wormhole to go back to a period before you created the wormhole. Even if the wormhole was previously created by someone else, it could only return you to the point at which you entered it. At any rate, traversable wormholes are probably impossible.

The slight caveat is that we cannot be 100 percent certain about any of this until we have a provable theory of quantum gravity. It's unlikely such a theory will drastically change the way we think the universe works, but we must acknowledge we cannot be sure until we have a theory that simultaneously accounts for general relatively and quantum effects.

The Future of Sci-Fi Technology

Some of these high-tech sci-fi technologies are credible, or even exist in some form today, but many are more magic than sci-fi tech. The bigger problem with many of these icons of science fiction is that they all represent an attempt at imagining contemporary people with largely conventional lives using advanced technology to accomplish familiar tasks or goals. We imagine, for example, a person who could be someone from 1970 living aboard a spaceship in the 2200s and using a holodeck the way we would use one.

But the people of the future, often in unpredictable ways, will be different, with different needs and customs. By the time we have a holodeck, interstellar space travel, or cryosleep, we may be genetically modified, AI-enhanced cyborgs partly existing in a virtual reality that seamlessly integrates with a digitized physical reality. The problems these technologies are trying to solve may not exist, and we may be facing new ones not relevant today.

All the sci-fi visions that spawned these awesome technologies fail because they imagine people implausibly like us using them—a necessary plot device because we have to relate to these futuristic characters, so we just put futuristic toys in the hands of knights, cowboys, explorers, villains, and so on. As a result, it distorts our image of what the future might actually be.

In this same way, we employ reverse engineering, trying to solve a contemporary problem with advanced technology, so we imagine a way for it to work. In reality as we advance toward that future technology, it will evolve in unpredictable ways, spawning off new applications, changing us and the world.

29. Regeneration/Immortality

"Death is an engineering problem."

So says the now-famous quote from computer scientist Bart Kosko. This has been the central belief for those who hold the view that there are no fundamental or theoretical reasons why people cannot be made immortal through technology. By "immortal" I mean essentially ageless with no upper limit on life span. An immortal biological being can still die through extreme trauma, poisoning, energy damage, or other means, but they won't die from aging alone.

The other view holds that no machine, whether biological or not, can be truly immortal. Systems simply don't work that way. They tend to break down, to build up waste, trauma, strain, or weaknesses, and eventually cannot maintain themselves. For example, genetic mutations are going to inevitably accumulate while repairs will be imperfect. At best we can achieve extreme life extension, but nothing approaching immortality.

There is also a third view, orthogonal to the first two, which holds that it doesn't matter whether we can achieve functional immortality—we shouldn't. Such a thing would be bad for our species and civilization, leading to a solidification of beliefs, power structures, and institutions. This would stifle ingenuity and progress, and lead to the stagnation of humankind, if not the collapse of civilization. On an individual level, there also would likely be a host of negative psychological effects.

These conflicting views are reflected in the classic works of the science fiction genre. Frank Herbert's *Dune* series imagines a future thousands of years from now, but where most people still live a recognizable life span of about eighty years. Most science fiction follows this pattern, I suspect because the societal impact of extreme life extension would radically alter their vision of the future in complex ways they would rather not deal with. By contrast, sci-fi in which humans are immortal through technology tend to focus on the immortality itself as a central theme, such as the classic *To Live Forever* by Jack Vance, and the more recent, *The Postmortal* by Drew Magary. Such books tend to focus on the societal problems provoked by immortality, rather than having it be an incidental fact of the world-building.

In trying to predict the future of technology and its implications, we need to determine how best to extrapolate life expectancy and life span into the future, not just if and when we cross the line into immortality.

The Science of Life Extension

Let me first deal with some terminology bookkeeping: Life expectancy is the statistical probability of living to a certain age starting at a specific age, assumed to be birth unless otherwise specified. In 2019, the global average life expectancy was 73.4 years. The country with the highest life expectancy was Hong Kong at 85.3 years, followed by Japan at 85 years.

Life span, rather, is how long a creature will live without any cause of premature death. We might think of this as how long a species lives before it dies of old age. So far for the human species, we have been able to increase our life expectancy simply by reducing the causes of premature death. Even reducing infant mortality increases life expectancy. However, it does not seem that we have extended the human life span at all. Even our distant ancestors could live to a ripe old age; they were just very unlikely to.

We therefore begin our quest with what seems like a simple

question but is actually a bit complex—what is the ultimate limit of the human life span? The oldest documented person to have ever lived was Jeanne Calment of France, who died in 1997 at the age of 122 years and 164 days. It's reasonable to argue that from a practical perspective, that's the upper limit. A 2021 study (Pyrkov et al.), however, took a biologically reductionist approach to this question, looking at markers of aging in the blood and extrapolating out to the point that life would simply not be feasible. They came up with an upper limit for the human life span of 150 years.

A 2018 study (Barbi et al.) took a different approach, looking at the risk of death by age. If dying is an inevitable consequence of aging, then the risk of dying should increase with age, but this study found that in the oldest living people (those 105 and older), the risk of death was flat. This plateau in death rate suggests no upper limit by itself, but of course we don't know if this will hold out past 122.

There is no question that with modern medicine and improved quality of life and safety, life expectancy increases. If everyone had perfect safety and health care, then we would expect life expectancy to approach the upper limit of life span. While this is enough to explain the increase in life expectancy, scientists also want to know if life span has been increasing, or to put it another way, has aging been slowing? Does improved nutrition, lifestyle, and health care actually reduce the rate at which our cells age? Is fifty really the new forty?

While this question is still not definitively resolved, a 2021 study (Colchero et al.) looked at humans and other primates and found that the rate of aging within a species is "invariant." In other words, aging and therefore life span are fixed.

Given what we know today, what changes can we expect in human life expectancy? This partly depends on social and political systems, not just technology. Things like pollution, income inequality, worker safety, and the availability of health care all play a huge role in life expectancy. We are not going to fix all of society's ills anytime soon, so there will likely continue to be a range, but we can extrapolate out life expectancy for the longest-lived developed nations.

Let's assume no disruptive technology, which seems to imply no change in the human life span. Medical technology continues to advance, with stem cell therapy, gene therapy, and at least basic nanotechnology, as well as all aspects of medicine continuing their incremental improvement. In this scenario, we would expect life expectancy to continue to slowly increase, into the 90s by the end of the century, and perhaps north of 100 by next century. The longest living people may break Calment's record of 122, and perhaps reach into the 130s.

It's hard to say what the ultimate limits of this increase will be, but a reasonable guess is a life span of 130 to 150 years with a life expectancy of 100 to 120 years. I also think it is reasonable to predict achieving these kinds of numbers between 100 and 300 years from now.

Strategies for Engineered Negligible Senescence

What if, however, we find ways of not only maximizing life expectancy but also extending the limits of the human life span? This requires not just medical care, but fundamentally altering our biology. That is the goal of researchers pursuing "strategies for engineered negligible senescence" (SENS), which means slowing the aging process to an imperceptible crawl. This term cleverly avoids the notion of "immortality" or perfection, but it gets close enough for practical purposes.

Perhaps the strongest proponent of SENS is biogerentologist Aubrey de Grey of University of Cambridge. His goal has been to identify all the specific biological mechanisms of aging and then find ways to fix them. This is a perfect example of the "engineering" approach to extreme life extension. The field is too vast to give a thorough treatment here, but one representative example is telomere length. The telomeres are the ends of chromosomes, like a cap at the end. Each time a cell reproduces and makes copies of its chromosomes, the telomeres shorten just a little.

Therefore, if we can find a way to lengthen the telomeres of every chromosome in all of our cells, that would reduce one important cause of aging. However, it is still not clear if telomere length is really a cause of aging or just a marker for aging. Lengthening telomeres may therefore be like a face-lift—it will not make you younger, but it will make you look younger (at the cellular level).

There are scores of other physiological things that happen as we age that could plausibly cause aging, like the buildup of waste material in cells and reduced effectiveness of DNA repair mechanisms. The idea is that if we can identify and fix all these specific things that cause aging, we could engineer ourselves younger.

Another way to accomplish the same goal is through regeneration, perhaps using stem cells or genetic manipulation. Proponents of the potential for immortality make an interesting point. Life itself is immortal. If you think about it, we are the descendants of a living cell from about 4 billion years ago. Multicellular life has been around for about 600 million years. On a cellular level, reproduction is a form of regeneration.

As discussed earlier in relation to stem cell therapy, there is a jellyfish, *Turritopsis dohrnii*, that is considered to be the only immortal animal. When it gets injured, it regenerates by essentially giving birth to itself. Stem cells are immortal in culture, and embryonic stem cells are biologically immortal. When cells differentiate into a specific adult type, they sacrifice their immortality in order to take on their specific functions, and also so that they stop reproducing and growing out of control.

If, therefore, you regenerated your liver from immortal stem cells, that would hit the reset button on the aging of your liver. You would not have to fix all the effects of aging in every liver cell, just grow a new liver.

The most extreme example of this would be to grow yourself an entirely new body, like the immortal jelly. The one limitation on this is your brain, because your brain is you. If you regenerated a new brain, then you would just be creating a clone, but one without your memories.

The brain, therefore, is the ultimate limiting factor on biological immortality. If we can keep the brain going through repair mechanisms, and rejuvenate our bodies through regeneration from stem cells, that is probably the closest we can get to biological immortality. How long can we keep our brains going? With conventional medicine, no more than the current life span. But with extreme tech like genetic alteration and nanotechnology, we can't know until we get there.

There is one path to regenerating the brain, however, if we do it a bit at a time. We can't just grow a new brain like you would a pancreas, but what if we were able to create neural stem cells to replace aging brain cells bit by bit? The brain's dynamic integrity would be maintained, while slowly regenerating.

It's very difficult to predict when such extreme life extension will become possible, and what its ultimate limits will be. This is extremely tricky technology, even if it sounds conceptually simple, and may take centuries to perfect. When it is perfected, what will that mean? Will these techniques allow humans to live for five hundred years, a thousand years, or even thousands of years?

Our Immortal Future

Let's say that in the centuries to come, through a variety of techniques, humans can live as long as they want (functional immortality) with reasonable quality of life. That last bit is important, as just living longer does not mean much if the quality is not there.

In fact, *Gulliver's Travels* (published in 1726), arguably the first science fiction book, dealt with this issue. During his voyages,

Gulliver visited the island of Luggnagg, where he encountered the struldbrugs, who are immortal. Unfortunately, their immortality did not come with eternal youth. (Remember this when making that carefully worded wish to a genie.) They kept aging into increasing decrepitude. As a result, Luggnagg citizens were considered "legally dead" at the age of eighty and could own no property, which was passed on to their heirs.

In our imagined future, people have indefinite youthful vigor and no expiration date on their life span. Would this be a good thing or a bad thing? There are vehement proponents on either side of this issue, and I feel this is one of those debates that is unresolvable and will be until we ever do achieve extreme life extension. Likely there will be good and bad things simultaneously, and the balance will depend on individual variability and collective policy.

The obvious benefit is that most people would not have to face the incremental disability that comes with old age. Ideally, everyone would live their life in perpetual good health and relative youth until ended by some sudden catastrophe. Alternatively, people could choose to end their life voluntarily, once they feel they have had enough of it. In this context, suicide would likely be considered a precious right, free of stigma.

Over the course of an extended lifetime, one could have multiple careers, and gain much experience and wisdom, to the benefit of society. It's even possible that societies where the average age was greater than 100 years might be more mature and look back at prior ages as one where the passions of youth ran wild.

Extended lifetimes would also likely provoke necessary legal changes. For example, a "lifetime" prison sentence would mean a very different thing to someone who could potentially live for a thousand years. Death sentences would become even more barbaric than many consider them today.

Other contracts may also have to be reimagined. "Till death do us part" may become unrealistic, even for the most romantically inclined. The same with a lifetime appointment. In fact, it may become a general rule of law that no contract, obligation, punishment, or even privilege can last more than a set amount of time. Everyone will have the right to renew themselves every century.

I doubt, however, anyone will want to surrender all their wealth, or even a significant portion of it every 100 years, but perhaps there will be "century taxes" or something similar. Further, more people may be able to accumulate wealth given a century or more to invest. What will happen when most are able to retire? Who will work?

Which leads us to the unavoidable downsides of extreme life spans. The young are responsible for a great deal of society's labor, ambition, creativity, and dynamism. Asimov dealt with this issue in his *Robot* series, with worlds occupied by the wealthy and long lived decaying from stagnation, while the young masses outcompeted them.

Psychologically it's hard to predict how people, on average, will face centuries of life. Will we generally become bored and cynical curmudgeons, unable to take joy in life? Will every movie be predictable, every art form derivative, and every politician transparent? And what about motivation? Without the clock ticking on life, perhaps we will endlessly procrastinate. Why do today what you can put off until next century? There will have to be a new psychology of the very old. We can only speculate what such psychologists will discover about the effects of extreme life extension.

Conceivably the most profound effect will be on population. If everyone is immortal, that leaves little room for children, a topic that is often very dark when addressed in fiction. In the series *Love Death + Robots*, the episode "Pop Squad" is set in a world of immortals where having children is illegal. A police force is established to essentially hunt down and summarily execute any illegal children and to arrest their parents. This is obviously soul-crushing for all involved, and the story explores those implications.

In the novel *Scythe*, author Neal Shusterman takes another

approach. An elite priesthood is granted the permission to cull the human population as needed, keeping the numbers controllable. They alone determine who should die, and their will should not be opposed.

However, even with a fully immortal population, the death rate would not be zero. There would still be accidental deaths, murders, and suicides. Still, it would be much lower than without biological immortality. This would allow for some births, which would have to be managed to avoid unlimited overpopulation, such as through a lottery every century to determine if you are allowed to have a child.

To fully imagine this future, we need to include the effects of other technologies. Perhaps people will move to settlements on the moon or Mars just so they can have children. It's possible we will turn to robot children. Or maybe people will be pushed into a virtual existence where they can have as many virtual offspring as they wish.

Regardless, it's clear that life with few or no children will be very different, and likely to have a large psychological impact on individuals and society. This problem may even partially take care of itself, with the inability to have children leading many people to decide they don't want to live past 200 or so. Some equilibrium may then emerge, balancing the benefits and detriments of life extension as a whole.

Speculating about the effects of biological immortality may be near impossible because humanity will have evolved to such a point by then that we cannot currently imagine ourselves in that future. We will be genetically modified cyborgs living in a world utterly transformed by our technology (that is, if we want to). I don't know how those future derivatives of humanity will feel about children. I only know how I do.

30. Uploading Consciousness/ The Matrix

"I know kung fu."

In discussing the brain-machine interface, we touched upon the implications of a biological brain living in a fully robotic body, or a virtual reality. The science fiction series *Altered Carbon* flips this around and imagines a digital mind in a biological body. Those with resources can have their minds transferred into a small disc-like computer called their "stack," which can then be inserted into a receptive body, which they call their "sleeve." You are the stack, and a sleeve is just a temporary thing you wear, like a jacket.

Is this a viable pathway to near immortality, especially if biological immortality does not work out? There are two components to this question. The first is easier to deal with—is a digital mind immortal? The answer is potentially, as long as the digital information can be maintained indefinitely.

Once your memories, personality, feelings, and everything that makes you *you* is fully recorded in a digital medium, that opens up lots of possibilities. You could exist in a purely digital universe, as part of that digital universe (not just hooked up to it). The experience would be as real as you existing and reading this book right now. You could interface with those on the outside with an avatar, like Max Headroom (although not so 1980s). This idea was explored in the fantastic *Black Mirror* episode "San Junipero," and more humorously in the TV series *Upload*.

You could also exist in the physical world, as in *Altered Carbon*, by having a brain-machine interface, but in reverse. Your digital self would take control of the meat puppet, and you would fully feel as if you were the biological body, just as you feel that you are your current body. If your current body is injured or wears out, just transfer to a new one.

There would be other options, such as being inserted into biological organisms other than unmodified humans (like in the movie *Avatar*). For example, we could then genetically engineer humans or other body types to be adapted to living in the sea. This might also be a method of adapting to settlements on other worlds. If you want to spend some time on Mars with low gravity and a thin atmosphere, no problem. Just insert your "stack" into a martian body. When you return to Earth, you can go back into your Earth-adapted body.

Some environments might be too extreme for biology, or there might be jobs that require some hard technology. In that case, just enter a specially designed robot. As previously stated, we may not need to fear a future robot apocalypse, because we will be those robots.

If you need to be in two places at once, no problem. Just make a copy of yourself, each with its own optimal sleeve. You can recombine the separate experiences later into one unified self. It may also be possible for the multiple versions of you to never actually separate but rather be connected wirelessly, sharing information in real time.

The possibility of death can also be further minimized by making regular backup copies of yourself. If there is an accident where your body is completely destroyed, including whatever medium contained your mind, then the backup copy is there waiting.

Clearly, in a technologically advanced civilization, there are advantages to being a digital entity. This brings us to the second and more difficult question: Is a digital copy of you really you? This is sometimes referred to as the "continuity problem."

From one perspective, one that I subscribe to, you cannot "upload your consciousness" into a computer or any medium. This is a misleading phrase, and makes it falsely seem like your mind is somehow moving from your brain to the computer. There is no *Freaky Friday* scenario. In reality, your mind is not a thing that can be moved. It is the functioning of your brain, and they cannot be separated. At most you are copying the information in your brain onto a computer, and that copy is just that, a copy.

Perhaps this doesn't matter. Let's say you are going to die anyway, at least now there is a version of you in the world. This might be a comfort to your loved ones, while you are forever in oblivion. It's also great for the digital you, but that is simply not you.

There is a related and complex philosophical question of whether information by itself can be considered conscious or self-aware. Is the backup copy of you a conscious being? Or is consciousness a combination of the information and the physical medium that processes that information?

Our biological brains are not divided into hardware and software. Our brains are wetware, with information, memories, and processing circuits all being the same thing. What this means is that a backup copy of your consciousness would not be conscious, but it would become so if it were encoded into hardware capable of generating consciousness.

We are then, however, back to the continuity problem of whether that backup copy transferred to a new medium would be you, or just another copy. I think the latter is most likely the case. But the lines here are getting quite blurry.

If, however, you are a continuity problem purist that will only accept a method that involves true continuity, is there a method of digital transfer that can work? Possibly.

I think the best method of digital transfer with maximum continuity involves a period of prolonged interfacing between the new digital medium and your biological brain. If, prior to transfer, you accept the premise that you are your brain, then we can't just get rid of the brain. But what if we used a BMI to connect your biological brain to a computer that itself was fully capable of artificial intelligence, although currently empty of any memories? This digital brain would function like a brain augmenter.

In essence you would become a hybrid consciousness between your biological brain and the computer brain, which are massively connected and networked together (just like the two hemispheres of your brain). The computer and the interface would be designed so that over time your brain would store memories on the computer part of itself, using it like a third hemisphere. The computer hemisphere may slowly take over more and more of the circuits of your brain.

Eventually, the biological hemispheres may become entirely redundant. In fact, if the digital component of your new brain is powerful enough, the biological components may be insignificant. You could test this, and what it would feel like to exist only in the digital hemisphere, by turning off the biological ones every night when you slept. In the deeper stages of sleep, only your digital brain would be active, and at some point you may stop noticing any difference. You are now ready to enter your fully digital life.

Some futurists have proposed that we could accomplish this much quicker by slowly replacing the neurons in the brain with artificial neurons. Each neuron would function like the biological one, an unnoticeable change, until eventually your entire brain was digital rather than biological. Would this maintain continuity? I admit, I'm not sure, but I like my method better.

In any case, we potentially have two methods for making digital human minds, either through some kind of slow migration if you want to maintain at least the possibility of continuity, or just copying if you don't care. Once this is achieved, part of humanity will be digital, which will have profound effects on our civilization.

Such a change will alter the very definition of what it means to be human. A physical body is no longer required, and the form such a body takes will become increasingly less important. It may not even matter if the body is biological or mechanical.

We could, in fact, occupy a body made of foglets—programmable matter. Our foglet body could take any form at will, adapting to the environment, situation, or task at hand. We can do this even if we still exist as a biological brain, but one that is interfaced with a digital component. As in the movie *Surrogates*, our frail bodies could be safely contained in a secure location, while we live out our physical lives through an advanced simulacrum.

Without the constraints of a purely biological brain, things like learning will also radically change. It might become a simple matter of downloading information or subroutines into your consciousness program. We may move seamlessly from a solo virtual mind to a group mind and to a large number of possible physical hosts. Our identity would be completely divorced from any physical form, which we could change as easily as changing clothes.

Space travel would also be much easier. If you need a copy of yourself in a distant system, you can just beam the information at the speed of light. It would still take years, but that's the fastest travel likely to be possible. If you need your physical continuity to travel to a distant star, then at least you won't need to drag a biological body along with it. During travel, you can simply go into sleep mode, making the journey seem instantaneous even if it took decades. Or you can spend the time in a virtual world.

Virtual worlds could also have variable time rates. If you need to get something done quickly, you can enter a virtual world that experiences time a hundred times faster than normal human time. We could use this method to speed up scientific research or creative projects.

On the other hand, a fully digital spacefaring civilization could dramatically slow down their collective experience of time. This would reduce needed processing power, which may be a critical resource for the civilization. Further, it would be another way to dramatically change space travel. If an entire civilization were operating at 1/1,000th the speed that humans naturally experience, then an interstellar trip that takes 100 years would feel like a month. You could get to Mars in a day.

Making relative cognitive speed a variable would have profound and far-reaching implications. Cognitive speed could be a matter of intense competition. Thinking more slowly than your competitor, or enemy, would be unacceptable. If a civilization sped up its relative cognition a thousand times, then the universe would become a much bigger place. Communicating at light speed would subjectively take twenty-one minutes to the moon, and eight days to Mars. Traveling to Mars, even in an advanced rocket that would take only a month, would subjectively feel like eighty-three years. Perhaps this is the ultimate answer to the Fermi paradox—interstellar distances would be a thousand times more daunting than they already are.

Even communicating on Earth would be tedious, with transmission delay time ranging from seconds to minutes. It would feel as if the entire natural world slowed down by a factor of a thousand, with everything looking frozen in time. How would civilization collectively decide where to set subjective time, and would different populations be effectively isolated from each other by their relative cognitive rates?

Digital consciousness is one of those possibilities that is so profound and transformative that it is difficult to imagine the implications. The reality is likely to be far different than any futurist can predict. At best we may envision some broad brushstroke possibilities.

It is also important to realize that of all the far future possibilities, this is perhaps the most plausible and likely to happen. There are no laws of physics in the way, no real breakthroughs or new advances in basic science required. Incremental advances in existing technology will get us there, possibly this century.

Therefore, any vision of the future must be seen through the lens of digital consciousness. All of the other technologies we discuss in this book, once we get even a few hundred years in the future, will be used at least in part by people who are no longer restricted to biological existence. This will impact space travel, our feelings about biological manipulation, and our relationship with all other technology. Once we achieve digital consciousness, we will have become our technology.

Conclusion

The future is now.

Science and technology have clearly transformed human civilization and will continue to do so in the future. In reviewing the advances that have brought us to where we are, an exciting reality has come into focus for me—the future is now. We are already living in the future, one that exceeds the exuberant imaginations of past futurists in many ways.

Robots are taking over manufacturing, creating unprecedented productivity. The world's store of information is accessible through a small device you can carry in your pocket, which you can also use to communicate with individuals, groups, and even the public. We have genetically engineered crops, advanced solar energy, electric cars, and metamaterials with properties previously thought impossible. Medicine likewise has advanced, with mRNA-based vaccines, monoclonal antibody therapy, early gene therapy, and brain-hacking treatments for neurological disease.

We are living in a future that would have blown away the science fiction writers and futurists of the past. As a species, we have made important progress. And yet, we are not living in a techno-utopia. We have inherited all the social ills, biases, bigotry, and conflicts of our ancestors. We are collectively wiser than our ancestors, but still have a long way to go.

Even technologically, we still must face numerous trade-offs and difficult choices. Those choices will shape our future. We therefore cannot predict the future of technology without partially predicting the future of humanity itself. In what ways will future humans be similar to or different from us today? What decisions will those future people make?

Some aspects of technology are easy to predict, at least in general terms. There are many technologies experiencing rather predictable steady incremental advances that will likely continue, as least for a time. Batteries will have greater energy density, computers will become more powerful, cancer survival will slowly increase, and robots will become more autonomous.

Other aspects of technology are essentially coin flips, subject to economic factors and the vagaries of public opinion. When will fusion power be practical? How much will we invest in human space travel versus robotic exploration? How tightly will we regulate genetic manipulation?

We can also make some predictions about large trends, even if we cannot know the details. It seems clear that we are moving in the direction of greater digitalization of technology, more of a direct connection between information and physical reality. Our world is increasingly becoming virtual, run not only by computers but also by powerful artificial intelligence algorithms.

Information itself is increasingly our most powerful resource, overriding all others. With genetic information, we are gaining more control over biology. With additive manufacturing and nanotechnology, we are increasingly able to translate our virtual world into physical reality.

Even humans will likely become more digital, more virtual, and more connected. Virtual/augmented reality and brain-machine interface technology will likely increasingly immerse us in a virtual world. The lines between biology and machines, between brains and computers, will not only be blurred but also may be obliterated.

Conversely, it's difficult to predict which hurdles will become deal-killers, rather than inconveniences. Some technologies seem forever delayed because certain obstacles have no practical solution. Perhaps we will never have a hydrogen revolution because it's simply too bulky to transport safely. Genetic manipulation of humans may encounter safety issues that delay its wide adoption for decades or even centuries. Fusion power may have to wait for advanced materials we cannot currently imagine.

Then there are the unknown unknowns, the unpredictable truly disruptive technologies. Pre-1960s futurists imagined an analog future, one that will never come to be. It is clear now that our future is digital, but this was entirely unpredicted before it actually happened. What will be the disruptive technologies of the future? We cannot know. We can only speculate.

Physicists may discover some aspect of the universe previously not hinted at, one that will enable us to slip through space at considerably faster-than-light speeds. We could also discover a source of energy that is currently unknown, rendering all predictions of our future energy infrastructure instantly obsolete. Perhaps we will even discover the right combination of genetic alterations to allow for full regeneration.

I am not holding my breath on any of these breakthroughs. But should they happen, these are advances that will improve our ability to play the game and will change the very game itself. Suddenly old trade-offs are gone, and we will face an entirely new set of possibilities.

Some game changers are perhaps more predictable, in that they are likely to happen eventually, although when and what the result will be are less clear. I file general artificial intelligence in this category. We know it's coming. It may be 50 years or 100 years, but at some point, we will be able to create a general artificial intelligence that is as smart as a human, and then soon after that, 100 million times as smart.

This is a nexus point. It's hard to imagine the world afterward because there are so many different possibilities. How much power and autonomy will we give such AIs? How will we use them, and is that even a meaningful question? To what extent will we merge with our AI, becoming hybrid technological and organic intelligences ourselves? How will society react?

We must also consider what technological booby traps may be lying in wait for us. We already are facing a big one with global warming, which will certainly affect our current and future decisions about technology. Some futurists worry that unchecked human population growth made possible by technology will begin to dominate our lives, an "overpopulation apocalypse" if you will, as in the movie *Soylent Green* (ironically set in 2022), and many other works of science fiction. Or perhaps population becomes a nonissue with a rise in quality of life and decrease in poverty.

With the leeway in how we project our society and technology into the future, you can either be an optimist or a pessimist. If history is any guide, reality will likely fall somewhere in the middle, with genuine advancement and improvement alongside persistent serious problems.

In the end, the most important variable is not technology itself but people. Our political, ethical, judicial, and professional institutions likely will have more to say about our future than just technological advance. Technology will not save us from ourselves. We can use our technology to create or destroy, free or enslave, to enlighten or control. These are the choices that will dominate our future.

Ultimately, technology is easier to predict than people, making the most important future advances those in critical thinking, philosophy, and even psychology. The people in the future will not be us. Who will they be?

The only thing that is certain about the future is that it will be different in ways we cannot currently imagine and will be occupied by people who think we are quaint and perhaps even barbaric. Things will be different that we didn't even know could be, or perhaps didn't even know were things.

But we will get there, slowly. We will craft the future one day at a time.

References

1. Futurism—Days of Future Passed

Irving, Richard, Gil Mellé, and Hal Mooney. *The Six Million Dollar Man*. USA, 1973.

Mann, Adam. "This is the Way the Universe Ends: Not with a Whimper, but a Bang." *Science*, August 11, 2020. https://www.science.org/content/article/way-universe-ends-not-whimper-bang.

Scott, Ridley, Vangelis & Vangelis. *Blade Runner*. USA, 1982.

Spielberg, Steven, John Williams, and John Neufeld, C. P. *Minority Report*. USA, 2002.

2. A Brief History of the Future

Asimov, Isaac. "Visit to the World's Fair of 2014." *New York Times*, August 16, 1964. https://archive.nytimes.com/www.nytimes.com/books/97/03/23/lifetimes/asi-v-fair.html.

"The City of the Future (1935)." https://www.youtube.com/watch?v=UZUMo_QYbB0.Driving.

"Despite Repeated Attempts, Turbine Cars Just Never Took Flight." March 16, 2018. https://driving.ca/auto-news/news/the-troubled-history-of-the-turbine-car.

"From 1956: A Future Vision of Driverless Cars." https://www.youtube.com/watch?v=F2iRDYnzwtk.

"Retro 1920's Future—What the Future Will Look Like!" https://www.youtube.com/watch?v=oXXSUzbKxZ4.

Saad, Lydia. "The '40-Hour' Workweek Is Actually Longer—by Seven Hours." Gallup. August 29, 2014. https://news.gallup.com/poll/175286/hour-workweek-actually-longer-seven-hours.aspx.

Ward, Marguerite. "A Brief History of the 8-Hour Workday, Which Changed How Americans Work." May 3, 2017. https://www.cnbc.com/2017/05/03/how-the-8-hour-workday-changed-how-americans-work.html.

"Year 1999 AD." https://www.youtube.com/watch?v=TAELQX7EvPo.

3. The Science of Futurism

Asimov, Isaac. *Foundation*. New York, NY: Gnome Press, 1951.

Bell, Wendel. *Foundations of Futures Studies: History, Purposes, and Knowledge*. Vol. 1. New York, NY: Routledge, 2003.

Cronkite, Walter. "Walter Cronkite in the Home Office of 2001 (1967)." https://www.youtube.com/watch?v=V6DSu3IfRlo.

Kurzweil, Ray. *The Age of Spiritual Machines: When Computers Exceed Human Intelligence.* New York, NY: Viking, 1999.

Samuel, Lawrence. "A Brief History of the Future." *Psychology Today*, January 27, 2020. https://www.psychologytoday.com/us/blog/future-trends/202001/brief-history-the-future.

Sherden, William. *The Fortune Sellers: The Big Business of Buying and Selling Predictions.* New York, NY: Wiley, 1999.

Twain, Mark. "From the 'London Times' of 1904." 1898. https://www.gutenberg.org/files/3251/3251-h/3251-h.htm#link2H_4_0009.

Vanderbilt, Tom. "Why Futurism Has a Cultural Blindspot." Nautilus. September 10, 2015. https://nautil.us/issue/28/2050/why-futurism-has-a-cultural-blindspot.

Watkins, John Elfreth, Jr. "What May Happen in the Next Hundred Years." *Women's Home Journal*, 1900. https://www.flickr.com/photos/jonbrown17/2571144135/sizes/o/in/photostream/.

4. Genetic Manipulation

AIEA. Plant Breeding and Genetics: http://www-naweb.iaea.org/nafa/pbg/.

Boyer, Herbert W., and Stanley N. Cohen. Science History Institute. https://www.sciencehistory.org/historical-profile/herbert-w-boyer-and-stanley-n-cohen. Human Genome Project. https://www.genome.gov/human-genome-project.

Cyranoski, David, and Heidi Ledford. "Genome-Edited Baby Claim Provokes International Outcry." *Nature News*, November 26, 2018.

Novella, Steven. CRISPR vs Talen. *Science-Based Medicine*. January 27, 2021. https://sciencebasedmedicine.org/crispr-vs-talen/.

Nunez, J. K. et al. "Genome-Wide Programmable Transcriptional Memory by CRISPR-Based Epigenome Editing." *Cell* 184, no. 9 (April 29, 2021).

Sanders, Robert. "FDA Approves First Test of CRISPR to Correct Genetic Defect Causing Sickle Cell Disease." March 30, 2021. https://news.berkeley.edu/2021/03/30/fda-approves-first-test-of-crispr-to-correct-genetic-defect-causing-sickle-cell-disease/.

5. Stem Cell Technology

"Estimated Number of Organ Transplantations Worldwide in 2019." Statista. 2019. https://www.statista.com/statistics/398645/global-estimation-of-organ-transplantations/.

Merkle, F., S. Ghosh, N. Kamitaki et al. "Human Pluripotent Stem Cells Recurrently Acquire and Expand Dominant Negative P53 Mutations." *Nature* 545 (2017): 229–233. https://doi.org/10.1038/nature22312.

"Organ, Eye and Tissue Donation Statistics." Donate Life. 2021. https://www.donatelife.net/statistics/.

Watts, G. "Georges Mathé." *Lancet* 376, no. 9753 (2010): 1640. https://doi.org/10.1016/s0140-6736(10)62088-0.

Zakrzewski, W., M. Dobrzyński, M. Szymonowicz et al. "Stem Cells: Past, Present, and Future." *Stem Cell Research & Therapy* 10, no. 68 (2019). https://doi.org/10.1186/s13287-019-1165-5.

6. Brain-Machine Interface

Clynes, Manfred, and Nathan Kline. "Cyborgs and Space." *Astronautics*. September 1960.

Koralek, A., X. Jin, J. Long II et al. "Corticostriatal Plasticity Is Necessary for Learning Intentional Neuroprosthetic Skills." *Nature* 483, (2012): 331–335. https://doi.org/10.1038/nature10845.

Obaid, Abdulmalik et al. "Massively Parallel Microwire Arrays Integrated with CMOS Chips for Neural Recording." *Science Advances* 6, no. 12 (March 2020): eaay2789.

Warneke, Brett et al. "Smart Dust: Communicating with a Cubic-Millimeter Computer." *Computer* 34 (2001): 44–51.

7. Robotics

Argotc, Linda, and Paul Goodman. "Investigating the Implementation of Robotics." The Robotics Institute Carnegie Mellon University, Carnegie-Mellon University, Feb. 1984, www.ri.cmu.edu.

Barron, J. P., and P. E. Easterling. "Hesiod." *The Cambridge History of Classical Literature: Greek Literature*. Cambridge, UK: Cambridge University Press, 1985.

Čapek, K., and W. Mann. *R.U.R. Rossum's Universal Robots: RUR* (*Rossumovi Univerzální Roboti)—A Fantastic Melodrama in Three Acts and an Epilogue*. Independently published, 2021.

Edwards, David. "Amazon Now Has 200,000 Robots Working in Its Warehouses." *Robotics & Automation News*. January 21, 2020. roboticsandautomationnews.com/2020/01/21/amazon-now-has-200000-robots-working-in-its-warehouses/28840.

Frumer, Yulia. "The Short, Strange Life of the First Friendly Robot." *IEEE Spectrum* (May 2020).

Garykmcd. "The Brain Center at Whipple's." *The Twilight Zone* (TV Episode 1964). IMDb. November 1967. www.imdb.com/title/tt0734633/?ref_=ttep_ep33.

"How Robots Change the World—What Automation Really Means for Jobs, Productivity and Regions." Oxford Economics. Accessed August 11, 2021. www.oxfordeconomics.com/recent-releases/how-robots-change-the-world.

"IFR Presents World Robotics Report 2020." IFR International Federation of Robotics. Accessed January 24, 2022. ifr.org/ifr-press-releases/news/record-2.7-million-robots-work-in-factories-around-the-globe.

Leonard, M. "Each Industrial Robot Displaces 1.6 Workers: Report." March 3, 2020. https://www.supplychaindive.com/news/industrial-robot-displaces-16-workers/573248/.

Mayor, A. *Gods and Robots: Myths, Machines, and Ancient Dreams of Technology*. Princeton, NJ: Princeton University Press, 2018a.

Mayor, A. *Gods and Robots: Myths, Machines, and Ancient Dreams of Technology*. Princeton, NJ: Princeton University Press, 2018b.

Mayor, A. "When Robot Assassins Hunted Down Their Own Makers in an Ancient Indian Legend." March 18, 2019. https://qz.com/india/1574936/robots-guarded-buddhas-relics-in-ancient-indian-mythology/.

McFadden, C. "The History of Robots: From the 400 BC Archytas to the Boston Dynamics' Robot Dog." July 8, 2020. https://interestingengineering.com/the-history-of-robots-from-the-400-bc-archytas-to-the-boston-dynamics-robot-dog.

Morris, Andrea. "Prediction: Sex Robots Are The Most Disruptive Technology We Didn't See Coming." *Forbes*, September 26, 2018. www.forbes.com/sites/andreamorris/2018/09/25/prediction-sex-robots-are-the-most-disruptive-technology-we-didnt-see-coming/?sh=1a60464d6a56.

Shashkevich, Alex. "Mythical Fantasies About Artificial Life." Stanford University. March 6, 2019. https://news.stanford.edu/2019/02/28/ancient-myths-reveal-early-fantasies-artificial-life/.

Smith, Aaron, and Janna Anderson. "AI, Robotics, and the Future of Jobs." Pew Research Center: Internet, Science & Tech. August 6, 2020. www.pewresearch.org/internet/2014/08/06/future-of-jobs.

8. Quantum Computing

Arute, F., K. Arya, R. Babbush et al. "Quantum Supremacy Using a Programmable Superconducting Processor." *Nature* 574 (2019): 505–510. https://doi.org/10.1038/s41586-019-1666-5.

Benioff, Paul. "The Computer as a Physical System: A Microscopic Quantum Mechanical Hamiltonian Model of Computers as Represented by Turing Machines." *Journal of Statistical Physics* 22, no. 5 (1980): 563–591. doi:10.1007/bf01011339.

"Beyond Qubits: Next Big Step to Scale up Quantum Computing." Science Daily. 2021. https://www.sciencedaily.com/releases/2021/02/210202113837.htm.

Cho, Adrian. "No Room For Error." Science. 2020. https://www.sciencemag.org/news/2020/07/biggest-flipping-challenge-quantum-computing.

Dr. Strangelove, Or, How I Learned to Stop Worrying and Love the Bomb. Culver City, CA: Columbia TriStar Home Entertainment, 2004.

Einstein, Albert. *The Born-Einstein Letters: Correspondence Between Albert Einstein, and Max and Hedwig Born from 1916 to 1955, Letter to Max Born, March 1948*. London: Walker & Company, 1971, 158.

Feynman, R. P. "Simulating physics with computers." *International Journal of Theoretical Physics* 21 (1982): 467–488. https://doi.org/10.1007/BF02650179.

Greig, Jonathan. "6 Experts Share Quantum Computing Predictions for 2021." Tech Republic. 2020. https://www.techrepublic.com/article/6-experts-share-quantum-computing-predictions-for-2021/.

Schumacher, Benjamin. "Quantum Coding." *Physical Review A* 51 (1995): 2738. https://journals.aps.org/pra/abstract/10.1103/PhysRevA.51.2738.

Steinhardt, Allan. "Radar in the Quantum Limit." Formerly DARPA's Chief Scientist, Fellow Answered June 30, 2016 "What could I do with a quantum computer that had one billion qubits?"

9. Artificial Intelligence

Branwen, G. GPT-3 Creative Fiction. June 19, 2020. https://www.gwern.net/GPT-3#why-deep-learning-will-never-truly-x.

Cellan-Jones, B. R. "Stephen Hawking Warns Artificial Intelligence Could End Mankind." BBC News. December 2, 2014. https://www.bbc.com/news/technology-30290540.

"Computer AI passes Turing Test in 'world first.'" BBC News. June 9, 2014. https://www.bbc.com/news/technology-27762088.

Good, I. J. "Speculations Concerning the First Ultraintelligent Machine." *Advances in Computers* 6 (1965): 31ff.

Kaplan, Andreas, and Haenlein, Michael. "Siri, Siri, in My Hand: Who's the Fairest in the Land? On the Interpretations, Illustrations, and Implications of Artificial Intelligence." *Business Horizons* 62, no. 1 (January 2019): 15–25. doi:10.1016/j.bushor.2018.08.004.

McCulloch, Warren S., and Walter Pitts. "A Logical Calculus of the Ideas Immanent in Nervous Activity." *Bulletin of Mathematical Biophysics* 5 (1943): 115–133. doi:10.1007/bf02478259.

Müller, Karsten, Jonathan Schaeffer, and Vladimir Kramnik. *Man vs. Machine: Challenging Human Supremacy at Chess.* Gardena, CA: Russell Enterprises, Inc., 2018.

Musk, Elon. "Blasting Off in Domestic Bliss." *New York Times*, 2020. https://www.nytimes.com/2020/07/25/style/elon-musk-maureen-dowd.html.

Turing, A. M. "I.—Computing Machinery and Intelligence." *Mind* LIX, no. 236 (October 1950): 433–460. https://doi.org/10.1093/mind/LIX.236.433.

10. Self-Driving Cars and Other Forms of Transportation

"Flying Cars Will Undermine Democracy and the Environment." American Progress. 2020. https://www.americanprogress.org/issues/economy/reports/2020/05/28/481148/flying-cars-will-undermine-democracy-environment/.

"Global EV Outlook 2021." IEA. 2021. https://www.iea.org/reports/global-ev-outlook-2021/trends-and-developments-in-electric-vehicle-markets.

Morando, Mark Mario, Qingyun Tian, Long T. Truong, and Hai L. Vu. "Studying the Safety Impact of Autonomous Vehicles Using Simulation-Based Surrogate Safety Measures." *Journal of Advanced Transportation*, 2018: 6135183, 22.04.2018.

"Road Traffic Injuries and Deaths—A Global Problem." CDC. 2020. https://www.cdc.gov/injury/features/global-road-safety/index.html.

"Ships Moved More Than 11 Billion Tonnes of Our Stuff Around the Globe Last Year, and It's Killing the Climate. This Week Is a Chance to Change." The Conversation. 2021. https://theconversation.com/ships-moved-more-than-11-billion-tonnes-of-our-stuff-around-the-globe-last-year-and-its-killing-the-climate-this-week-is-a-chance-to-change-150078.

Vaucher, Jean. "History of Ships Prehistoric Craft." Umontreal. 2014. http://www.iro.umontreal.ca/~vaucher/History/Ships/Prehistoric_Craft/index.html.

"What Happened to Blimps?" *Global Herald*, 2021. https://theglobalherald.com/news/what-happened-to-blimps/.

"World Vehicle Population Rose 4.6% in 2016." Wards Intelligence. 2017. https://wardsintelligence.informa.com/WI058630/World-Vehicle-Population-Rose-46-in-2016.

11. Two-Dimensional Materials and the Stuff the Future Will Be Made Of

"Concrete Needs to Lose Its Colossal Carbon Footprint." *Nature* 597 (2021): 593–594. https://www.nature.com/articles/d41586-021-02612-5.

"Global Crude Steel Output Increases by 4.6% in 2018." Worldsteel Association. 2019. https://www.worldsteel.org/media-centre/press-releases/2019/Global-crude-steel-output-increases-by-4.6-in-2018.html.

"The Global Natural Stone Market Was Valued at $35,120.1 Million in 2018, and Is Projected to Reach $48,068.4 Million by 2026, Growing at a CAGR of 3.9% from 2019 to 2026." GlobeNewswire. 2020. https://www.globenewswire.com/news-release/2020/01/16/1971569/0/en/The-global-natural-stone-market-was-valued-at-35-120-1-million-in-2018-and-is-projected-to-reach-48-068-4-million-by-2026-growing-at-a-CAGR-of-3-9-from-2019-to-2026.html.

Harmand, S., J. Lewis, C. Feibel et al. "3.3-Million-Year-Old Stone Tools from Lomekwi 3, West Turkana, Kenya." *Nature* 521 (2015): 310–315. https://doi.org/10.1038/nature14464.

12. Virtual/Augmented/Mixed Reality

"The Cinema of the Future." Mycours. 1955. http://mycours.es/gamedesign2016/files/2016/09/sensorama-the-cinema-of-future-morton.pdf.

"Google Glass Advice: How to Avoid Being a Glasshole." Guardian. 2014. https://www.theguardian.com/technology/2014/feb/19/google-glass-advice-smartglasses-glasshole.

"Mark in the Metaverse." The Verge. 2021. https://www.theverge.com/22588022/mark-zuckerberg-facebook-ceo-metaverse-interview.

"Worldwide Spending on Augmented and Virtual Reality Forecast." IDC. 2020. https://www.idc.com/getdoc.jsp?containerId=prUS47012020.

13. Wearable Technology

"The Invention of Spectacles | Encyclopedia.Com." Encyclopedia.com. Accessed January 24, 2022. www.encyclopedia.com/science/encyclopedias-almanacs-transcripts-and-maps/invention-spectacles.

14. Additive Manufacturing

Alexandrea P. "The Complete Guide to Fused Deposition Modeling (FDM) in 3D Printing." 3D Natives. September 3, 2019. www.3dnatives.com/en/fused-deposition-modeling100420174.

Balter, Michael. "World's Oldest Stone Tools Discovered in Kenya." American Association for the Advancement of Science. April 14, 2015. www.science.org/content/article/world-s-oldest-stone-tools-discovered-kenya.

"Bone Tools." The Smithsonian Institution's Human Origins Program. June 25, 2020. humanorigins.si.edu/evidence/behavior/getting-food/bone-tools.

"Chuck Hull Invents Stereolithography or 3D Printing and Produces the First Commercial 3D Printer: History of Information." HistoryofInformation.com. Accessed January 25, 2022. www.historyofinformation.com/detail.php?id=3864.

"François Willème Invents Photosculpture: Early 3D Imaging: History of Information." Accessed January 25, 2022. www.historyofinformation.com/detail.php?id=3876.

Gaget, Lucie. "3D Printing Creators: Meet Jean Claude André." 3D Printing Blog: Tutorials, News, Trends and Resources | Sculpteo. October 10, 2018. www.sculpteo.com/blog/2018/10/10/interview-meet-one-of-the-3d-printing-creators-jean-claude-andre.

"Injection Molding Market Size, Share and Growth Analysis Report." Bcc Research. May 2021. https://www.bccresearch.com/market-research/plastics/injection-molding-global-markets-and-technologies.html.

Lawton, C. "The World's First Castings." The C.A. Lawton Co. March 2, 2021. calawton.com/the-worlds-first-castings.

Lonjon, Capucine. "Discover the History of 3D Printer." 3D Printing Blog: Tutorials, News, Trends and Resources | Sculpteo. March 1, 2017. www.sculpteo.com/blog/2017/03/01/whos-behind-the-three-main-3d-printing-technologies.

Marchant, Jo. "A Journey to the Oldest Cave Paintings in the World." *Smithsonian Magazine*, January 6, 2016. www.smithsonianmag.com/history/journey-oldest-cave-paintings-world-180957685.

"A Record of Firsts—Wake Forest Institute for Regenerative Medicine." Wake Forest School of Medicine. Accessed January 25, 2022. school.wakehealth.edu/Research/Institutes-and-Centers/Wake-Forest-Institute-for-Regenerative-Medicine/Research/A-Record-of-Firsts.

Ślusarczyk, Paweł. "Carl Deckard—Father of the SLS Method and One of the Pioneers of 3D Printing Technologies, Died. . . ." 3D Printing Center. January 12, 2020. 3dprintingcenter.net/carl-deckard-father-of-the-sls-method-and-one-of-the-pioneers-of-3d-printing-technologies-died.

Statista. "3D Printing Industry—Worldwide Market Size 2020–2026." Statista. October 8, 2021. www.statista.com/statistics/315386/global-market-for-3d-printers.

"What Is a Lathe Machine? History, Parts, and Operation." Bright Hub Engineering. December 12, 2009. www.brighthubengineering.com/manufacturing-technology/59033-what-is-a-lathe-machine-history-parts-and-operation.

Williams, Nancy. "The Invention of Injection Molding." Fimor North America | Harkness Industries. February 27, 2017. harknessindustries.com/invention-injection-molding-2.

15. Powering Our Future

"Archaeologists Find Earliest Evidence of Humans Cooking With Fire." *Discover Magazine*, 2013. https://www.discovermagazine.com/the-sciences/archaeologists-find-earliest-evidence-of-humans-cooking-with-fire.

Berna, F. et al. Microstratigraphic Evidence of in Situ Fire in the Acheulean Strata of Wonderwerk Cave, Northern Cape Province, South Africa. Proceedings of the National Academy of Sciences, 109, no. 20 (2012): E1215–E1220. https://doi.org/10.1073/pnas.1117620109.

"Energy." Economist Intelligence. 2021. https://www.eiu.com/industry/energy.

England, P. C., P. Molnar, and F. M. Richter. "Kelvin, Perry and the Age of the Earth." *American Scientist* 95, no. 4 (2007): 342. https://doi.org/10.1511/2007.66.3755.

"IMF Estimates Global Fossil Fuel Subsidies at $8.1 Trillion as UN Urges Green Energy Push." Helenic Shipping News. 2021. https://www.hellenicshippingnews.com/imf-estimates-global-fossil-fuel-subsidies-at-8-1-trillion-as-un-urges-green-energy-push/.

KamLAND Collaboration. "Partial Radiogenic Heat Model for Earth Revealed by Geoneutrino Measurements." *Nature Geoscience* 4, (2011): 647–651. https://doi.org/10.1038/ngeo1205.

"A Look at Agricultural Productivity Growth in the United States, 1948–2017." USDA. 2021. https://www.usda.gov/media/blog/2020/03/05/look-agricultural-productivity-growth-united-states-1948-2017.

"Lost in Transmission: How Much Electricity Disappears between a Power Plant and Your Plug?" Insider Energy. 2015. http://insideenergy.org/2015/11/06/lost-in-transmission-how-much-electricity-disappears-between-a-power-plant-and-your-plug/.

"Smarter Use of Nuclear Waste." Scientific American. 2009. https://www.scientificamerican.com/article/smarter-use-of-nuclear-waste/.

"Space-Based Solar Power." Energy.gov. 2014. https://www.energy.gov/articles/space-based-solar-power.

"Tidal Power—U.S. Energy Information Administration (EIA)." Energy Information Association. 2021. https://www.eia.gov/energyexplained/hydropower/tidal-power.php.

"Who Discovered Electricity?" Universe Today. 2014. https://www.universetoday.com/82402/who-discovered-electricity/.

16. Fusion

"Advantages of Fusion." ITER. Accessed January 25, 2022. www.iter.org/sci/Fusion.

Arnoux, Robert. "Who Invented Fusion?" ITER. February 12, 2014. www.iter.org/newsline/-/1836.

Ball, P. "Laser Fusion Experiment Extracts Net Energy from Fuel." *Nature*, 2014. https://doi.org/10.1038/nature.2014.14710.

"The Birth of the Laser and ICF." Lawrence Livermore National Laboratory. Accessed January 26, 2022. www.llnl.gov/archives/1960s/lasers.

Clery, Daniel. "The Bizarre Reactor That Might Save Nuclear Fusion." Science.org. October 21, 2015. www.science.org/content/article/bizarre-reactor-might-save-nuclear-fusion.

"DOE Explains . . . Tokamaks." Energy.gov. Accessed January 25, 2022. www.energy.gov/science/doe-explainstokamaks.

Mott, Vallerie. "Isotopes of Hydrogen | Introduction to Chemistry." Courses.Lumenlearning.com. Accessed January 25, 2022. courses.lumenlearning.com/introchem/chapter/isotopes-of-hydrogen.

Paisner, J. A., and J. R. Murray. "National Ignition Facility for Inertial Confinement Fusion." OSTI.gov. October 8, 1997. www.osti.gov/biblio/631098.

"Physics of Uranium and Nuclear Energy—World Nuclear Association." World-Nuclear.Org, Nov. 2020. www.world-nuclear.org/information-library/nuclear-fuel-cycle/introduction/physics-of-nuclear-energy.aspx.

Power. "Fusion Power: Watching, Waiting, as Research Continues." *POWER Magazine*. December 3, 2018. www.powermag.com/fusion-power-watching-waiting-as-research-continues.

"Uranium Enrichment | Enrichment of Uranium—World Nuclear Association." World-Nuclear.org. September 2020. www.world-nuclear.org/information-library/nuclear-fuel-cycle/conversion-enrichment-and-fabrication/uranium-enrichment.aspx.

"Uranium Quick Facts." Depleted UF6. Accessed January 25, 2022. web.evs.anl.gov/uranium/guide/facts.

17. Mature Nanotechnology

Drexler, E. *Engines of Creation: The Coming Era of Nanotechnology*. Anchor Library of Science, 1987.

Feynman, R. P. "There's Plenty of Room at the Bottom." Reson 16, no. 890 (2011). https://doi.org/10.1007/s12045-011-0109-x.

18. Synthetic Life

Gibson, D. G. et al. "Creation of a Bacterial Cell Controlled by a Chemically Synthesized Genome." *Science* 329, no. 5987 (2010): 52–56. https://doi.org/10.1126/science.1190719.

Hutchison, Clyde A., III et al. "Design and Synthesis of a Minimal Bacterial Genome." *Science* 351, no. 6280 (March 2016).

Malyshev, Denis A. et al. "Romesberg: A Semi-Synthetic Organism with an Expanded Genetic Alphabet." *Nature* 509 (May 2014): 385–388.

Scott, R. *Alien*. London: Twentieth Century Fox, 1979.

19. Room-Temperature Superconductors

Bardeen, J., L. N. Cooper, and J. R. Schrieffer. "Theory of Superconductivity." *Physical Review*, 108, no. 5 (1957): 1175–1204. https://doi.org/10.1103/physrev.108.1175.

Hutchison, Clyde A., III et al. "Design and Synthesis of a Minimal Bacterial Genome." *Science* 351, no. 6280 (March 25, 2016). doi: 10.1126/science.aad6253. Erratum in: *ACS Chemical Biology* 11, no. 5 (May 20, 2016):1463. PMID: 27013737.

Koot, Martijn, and Fons Wijnhoven. "Usage Impact on Data Center Electricity Needs: A System Dynamic Forecasting Model." Science Direct. June 1, 2021. www.sciencedirect.com/science/article/pii/S0306261921003019.

"Nobel Prizes 2021." NobelPrize.org. Accessed January 26, 2022. www.nobelprize.org/prizes/physics/1972/summary.

O'Neill, M. "Prototype Microprocessor Developed Using Superconductors—80 Times More Energy Efficient." SciTechDaily. January 3, 2021. https://scitechdaily.com/prototype-microprocessor-developed-using-superconductors-80-times-more-energy-efficient/.

Snider, Elliot. "Superconductivity Warms Up." Nature Electronics. November 17, 2020. https://www.nature.com/articles/s41928-020-00507-3.

Snider, Elliot et al. "Room-Temperature Superconductivity in a Carbonaceous Sulfur Hydride." *Nature* 586 (October 15, 2020).

Strickland, J. "How Much Energy Does the Internet Use?" HowStuffWorks. July 27, 2020. https://computer.howstuffworks.com/internet/basics/how-much-energy-does-internet-use.htm.

van Delft, Dirk, and Peter Kes. "The Discovery of Superconductivity." *Physics Today* 63 (January 9, 2010). https://physicstoday.scitation.org/doi/10.1063/1.3490499.

van Delft, Dirk, and Peter Kes. "The Discovery of Superconductivity." Europhysics news.org. Accessed January 26, 2022. www.europhysicsnews.org/articles/epn/pdf/2011/01/epn2011421p21.pdf.

20. Space Elevators

Artsutanov, Y. "To the Cosmos by Electric Train." 1960. liftport.com. Young Person's Pravda.

"The Orbital Tower: A Spacecraft Launcher Using the Earth's Rotational Energy." U.S. Air Force Flight Dynamics Laboratory. 1975. http://www.star-tech-inc.com/papers/tower/tower.pdf.

Tsiolkovsky, K. E. "Speculations about earth and sky and on vesta." Moscow, Izdvo AN SSSR, 1959 (first published in 1895).

21. Nuclear-Thermal Propulsion and Other Advanced Rockets

Magee, J. G., Jr. "High Flight." Arlingtoncemetery.net. 1941. http://www.arlingtoncemetery.net/highflig.htm.

"Why Chemical Rockets and Interstellar Travel Don't Mix." Scientific American. 2017. https://blogs.scientificamerican.com/life-unbounded/why-chemical-rockets-and-interstellar-travel-dont-mix/.

22. Solar Sails and Laser Propulsion

Bussard, Robert W. "Galactic Matter and Interstellar Flight." Acta Astronautica 6 (1960): 179–195.

Chernov, D. *Man-Made Catastrophes and Risk Information Concealment: Case Studies of Major Disasters and Human Fallibility*. Zurich: Springer, 2016.

Marx, G. "Interstellar Vehicle Propelled by Laser Beam," *Nature* 211 (July 1966): 22–23.

Penoyre, Z., and E. Sandford. "The Spaceline: A Practical Space Elevator Alternative Achievable with Current Technology." arXiv:1908.09339 [astro-ph.IM] (2019).

Pettit, D. "The Tyranny of the Rocket Equation." NASA. 2012. https://www.nasa.gov/mission_pages/station/expeditions/expedition30/tryanny.html.

Schattschneider, P., and A. Jackson. "The Fishback Ramjet Revisited." ScienceDirect. February 1, 2022. https://www.sciencedirect.com/science/article/pii/S0094576521005804#!.

23. Space Settlements

Appelbaum, Joseph, and Dennis J. Flood. "Solar Radiation on Mars." Ntrs.Nasa.Gov. November 1989. ntrs.nasa.gov/api/citations/19890018252/downloads/19890018252.pdf.

Cain, Fraser. "What Is a Space Elevator?" Universe Today. October 10, 2013. www.universetoday.com/105441/what-is-a-space-elevator.

Clément, Gilles et al. "History of Artificial Gravity." Ntrs.Nasa.gov. Accessed January 26, 2022. ntrs.nasa.gov/api/citations/20070001009/downloads/20070001009.pdf.

David, Leonard. "Living Underground on the Moon: How Lava Tubes Could Aid Lunar Colonization." Space.com. July 30, 2019. www.space.com/moon-colonists-lunar-lava-tubes.html.

Howell, Elizabeth. "Axiom's 1st Private Crew Launch to Space Station Delayed to March." Space.com. January 20, 2022. www.space.com/axiom-1-launch-delay-march-2022.

Howell, Elizabeth. "International Space Station: Facts, History and Tracking." Space.com. October 13, 2021. www.space.com/16748-international-space-station.html.

Mathewson, Samantha. "How Recycled Astronaut Pee Boosts Chances for Future Deep-Space Travel." Space.com. November 16, 2016. www.space.com/34688-recycled-astronaut-pee-boosts-deep-space-travel.html.

National Space Society. "The Colonization of Space—Gerard K. O'Neill, Physics Today, 1974." National Space Society. March 29, 2018. space.nss.org/the-colonization-of-space-gerard-k-o-neill-physics-today-1974.

Rundback, Barbara. "The Stanford Torus as a Vision of the Future." The Rockwell Center for American Visual Studies. November 14, 2016. rockwellcenter.org/student-research/the-stanford-torus-as-a-vision-of-the-future.

Salisbury, F. B., J. I. Gitelson, and G. M. Lisovsky. "Bios-3: Siberian Experiments in Bioregenerative Life Support." *Bioscience* 47, no. 9 (October 1997): 575–85. PMID: 11540303.

Sarbu, Ioan, and Calin Sebarchievici. "Solar Radiation—an Overview | ScienceDirect Topics." ScienceDirect. 2017. www.sciencedirect.com/topics/engineering/solar-radiation.

Spry, Jeff. "Company Plans to Start Building Private Voyager Space Station with Artificial Gravity in 2025." Space.com. February 25, 2021. www.space.com/orbital-assembly-voyager-space-station-artificial-gravity-2025.

Sutter, Paul. "Lost in Space without a Spacesuit? Here's What Would Happen (Podcast)." Space.com. July 28, 2015. www.space.com/30066-what-happens-to-unprotected-body-in-outer-space.html.

24. Terraforming Other Worlds

"Altitude Physiology." High Altitude Doctor. Accessed January 29, 2022. http://www.altitudemedicine.org/physiology.

Buis, Alan. "Earth's Magnetosphere: Protecting Our Planet from Harmful Space Energy." NASA. November 16, 2021. https://climate.nasa.gov/news/3105/earths-magnetosphere-protecting-our-planet-from-harmful-space-energy/.

Hughes, David Y., and Harry M. Geduld. *A Critical Edition of The War of the Worlds: H.G. Wells's Scientific Romance*. Bloomington and Indianapolis: Indiana University Press, 1993.

Jenkins, D. R. "Dressing for Altitude." NASA. 2012. https://www.nasa.gov/pdf/683215main_DressingAltitude-ebook.pdf.

Mehta, Jatan. "Can We Make Mars Earth-Like Through Terraforming?" Planetary Society. April 19, 2021. https://www.planetary.org/articles/can-we-make-mars-earth-like-through-terraforming.

Space.com Staff. "17 Billion Earth-Size Alien Planets Inhabit Milky Way." Space.com. January 7, 2013. https://www.space.com/19157-billions-earth-size-alien-planets-aas221.html.

Steigerwald, B. "Mars Terraforming Not Possible Using Present-Day Technology." NASA. July 30, 2018. https://www.nasa.gov/press-release/goddard/2018/mars-terraforming/.

"Venus." NASA. (n.d.). Accessed January 29, 2022. https://solarsystem.nasa.gov/planets/venus/overview/.

Vilekar, S. A. "Performance Evaluation of Staged Bosch Process for CO2 Reduction to Produce Life Support Consumables." NASA. (n.d.). Accessed January 29, 2022. https://ntrs.nasa.gov/api/citations/20120015344/downloads/20120015344.pdf.

25. Cold Fusion and Free Energy

Ball, P. "Lessons from Cold Fusion, 30 Years On." Nature. May 27, 2019. https://www.nature.com/articles/d41586-019-01673-x.

Gibney, E. "Google Revives Controversial Cold-Fusion Experiments." Nature. May 27, 2019. https://www.nature.com/articles/d41586-019-01683-9.

Scientific American. "FOLLOW-UP: What Is the "Zero-Point Energy" (or 'Vacuum Energy') in Quantum Physics? Is It Really Possible That We Could Harness This Energy?" Scientific American. August 18, 1997. https://www.scientificamerican.com/article/follow-up-what-is-the-zer/.

26. Faster Than Light (FTL) Travel/Communication

American Physical Society. "Travel through Wormholes Is Possible, but Slow." ScienceDaily. April 15, 2019. Accessed January 25, 2022. www.sciencedaily.com/releases/2019/04/190415090853.htm.

Barr, S. "Folding Space—AAP. Ask A Physicist." 2015. https://wiki.physics.udel.edu/AAP/Folding_space.

Obousy, Richard K., and Gerald Cleaver. "Putting the Warp into Warp Drive." Cornell University. 2008. https://arxiv.org/abs/0807.1957v2.

27. Artificial Gravity/Antigravity

Borchert, M. J. et al. "A 16-Parts-Per-Trillion Measurement of the Antiproton-to-Proton Charge-Mass Ratio." *Nature* 601, no. 7891 (2022): 53–57. doi:10.1038/s41586-021-04203-w.

Motl, Luboš. "Is It Theoretically Possible to Shield Gravitational Fields or Waves?" Stack Exchange. 2011. https://physics.stackexchange.com/q/2809.

Wheeler, J. A. *Geons, Black Holes and Quantum Foam*. New York, NY: W. W. Norton & Company, 2000.

28. Transporters, Tractor Beams, Lightsabers, and Other Sci-Fi Gadgets

Haskin, B. *War of the Worlds*. Paramount Studios, 1953.

29. Regeneration/Immortality

Alexander, Vaiserman, and Dmytro Krasnienkov. "Telomere Length as a Marker of Biological Age: State-of-the-Art, Open Issues, and Future Perspectives." *Frontiers in Genetics* (January 2021).

Barbi, E. et al. "The Plateau of Human Mortality: Demography of Longevity Pioneers." *Science* 360 (2018): 1459–1461.

Colchero, F. et al. "The long Lives of Primates and the 'Invariant Rate of Ageing' Hypothesis." *Nature Communications* 12, no. 3666 (2021). https://doi.org/10.1038/s41467-021-23894-3.

Pyrkov, T. V. et al. "Longitudinal Analysis of Blood Markers Reveals Progressive Loss of Resilience and Predicts Human Lifespan Limit." *Nature Communications* 12, no. 2765 (2021). https://doi.org/10.1038/s41467-021-23014-1.

30. Uploading Consciousness/The Matrix

Morgan, R. *Altered Carbon*. London: Gollancz, 2018.

Conclusion

Fleischer, Richard, and Fred Myrow. *Soylent Green*. USA, 1973.

Index